New Ideas in Astronomy

For my little darling Elena,
whose sense of curiosity and wonder
about the world around them
reinspired my own.

Preface

This book is a compilation of nine articles written over an eighteen month period. Each chapter explores an astronomy topic that isn't discussed frequently in literature. The first four chapters cover topics in cosmology and collectively describe a vision of the universe that is sustainable, dynamic, infinite, and compatible with some versions of fractal cosmology. The fifth, sixth and seventh chapters explore astrophysical processes of our solar system's past and present that can be extended and applied to planetary systems in general. The last two chapters are related to our own planet's climate during the Quaternary Epoch. To the reader, may this book be a thought provoking and enjoyable read.

It's important to note something about this book's referencing. References are mentioned within the text and are placed at the end of sentences to improve the flow. The references do not always support the gestalt of their sentence, but merely show where factual information comes from such as measurements and relations between variables. References are summarized at the end of each chapter in APA bibliographical form.

Table of Contents

Black holes recycle matter and make light elements abundant in the universe

Abstract

Black holes at the center of Active Galactic Nuclei take in accretionary material, accelerate it, break it apart, and emit protons and electrons from their polar regions at very large fractions of the speed of light. This cosmic recycling process allows the universe as a whole to replenish its supply of hydrogen, which counteracts the production of heavier elements in the processes of stellar evolution and supernovae. This is important because if astrophysical jets are replenishing the universe's supply of hydrogen, then the universe could be capable of sustaining its abundance of light elements indefinitely. In that case, the Big Bang Theory would not be necessary for explaining the abundance of light elements in the universe.

Introduction

In 1948, Ralph Alpher of George Washington University predicted an abundance of light elements in the universe based on the Big Bang Theory (Alpher et al 1948). He predicted that hydrogen, helium and lithium should make up a very high proportion of atoms in the universe due to nucleosynthesis in the aftermath of the Big Bang and the finite age of the universe. Astronomical measurements of the cosmic helium abundance in the 1960s confirmed his basic prediction and propelled the Big Bang Theory ahead of its main competitor, the Steady State Theory (Hoyle and Tayler 1964, Wagoner et al 1967, Reeves 1974, Boesgaard and Steigman 1985). Calculations for the predicted ratios of light elements were refined in the late 1960s (Peebles 1966). Theoretical abundance curves for light elements and their observed values are displayed in figure 1A. Due to basic proton-proton fusion reactions, the abundance of ^{4}He tends to increase with increasing interactions between baryons while the abundances of ^{2}H and ^{3}He decrease.

For more than half a century, the abundance of light elements in the universe has been one the three main pillars of support for the Big Bang Theory alongside Hubble's Law and the Cosmic Microwave Background Radiation. Yet in the 21st century, discrepancies between measured cosmic abundance ratios and their theoretically predicted values prompted some astronomers to suggest that Big Bang nucleosynthesis is neither the necessary or optimal explanation for the abundance of light elements (Lopez-Corredoira 2022).

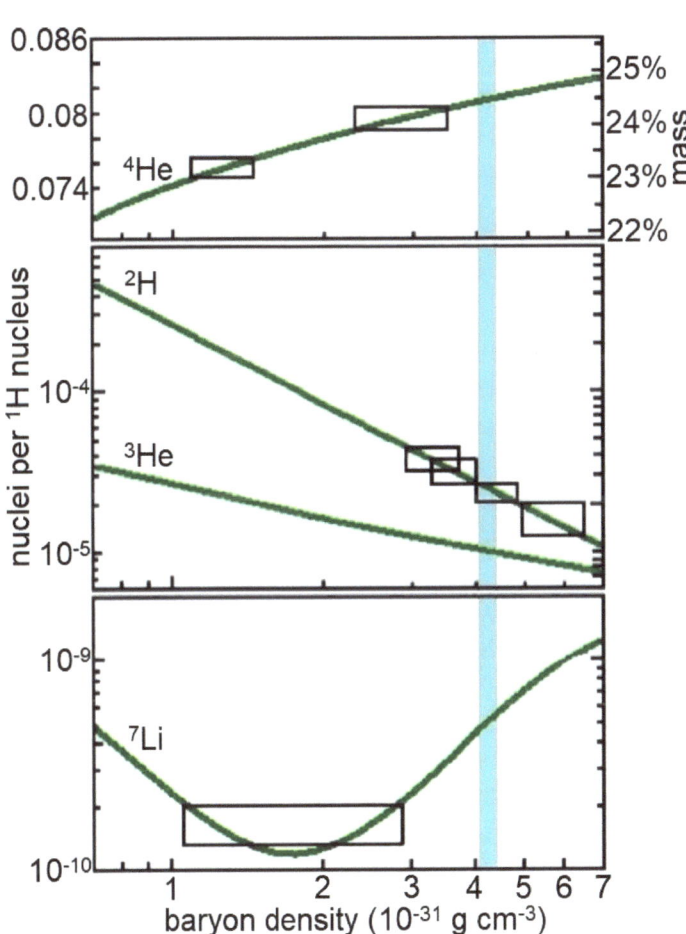

Figure 1A: Theoretical nuclear abundance curves versus observational measurements for ^{4}He, ^{2}H, ^{3}He and ^{7}Li (Kenath and Sivaram 2008). The green lines are the theoretical curves. The black rectangles represent the range of observed values within 2σ. The blue vertical band represents the range observed by WMAP.

Generation of Heavy Elements

Hydrogen, the universe's most abundant element, makes up about 74% of its mass, followed by helium at 24% (Croswell 1995). The remaining 2% is comprised of metals, which to astronomers actually means all elements larger than helium. Based on a measured baryon density of 4.2×10^{-31} g cm^{-3}, cosmic abundance ratios predicted by current nuclear reaction simulations are about 0.25 for ^{4}He:H, 10^{-3} for ^{2}H:H, 10^{-4} for ^{3}He:H and 10^{-9} for ^{7}Li:H (Kolb and Turner 2018, Spergel et al 2003). These values are in good agreement with the observational data shown in figure 1A, although the observed ^{7}Li abundance is higher than the value predicted by current Big Bang models by about a factor of 3 (Cyburt et al 2008, Fields 2011).

Stars are born in clouds of interstellar gas that are mainly composed of light elements and when a stellar engine ignites, it begins converting hydrogen nuclei into heavier elements via nuclear fusion. This in turn is how stars produce all of their heat and light. Over millions of years, stars transform large amounts of hydrogen into heavier elements through proton-proton chain reactions, the triple-α process, the CNO cycle and a multitude of less common nuclear reactions (Clayton 1968). When a star exhausts the supply of hydrogen in its core, the core starts to collapse until it gains enough temperature and pressure to start fusing helium nuclei. Once the supply of helium is exhausted, the star begins fusing heavier elements. This rapid stage at the end of a star's life is characterized by an intense temperature spike, a gravitational collapse in the core, a shock wave that initiates the generation of even heavier elements, and a supernova explosion (Woosley et al 1973).

Stellar processes generally transform light elements into heavier ones. Without a method for replenishing hydrogen, the universe would eventually be dominated by metals. This is certainly not the case in our universe.

Astrophysical Jets

Enter black holes, an idea initially explored by John Michell and Pierre-Simon Laplace in the late 18th century and theorized rigorously in the mathematical calculations of Karl Schwarzschild and David Finkelstein (Schwarzschild 1916, Finkelstein 1958, Montgomery 2009). By the 1960s, black holes were considered a natural consequence of Einstein's theory of general relativity. The first observational evidence for one was confirmed in 1971 with the discovery Cygnus X-1, a galactic X-ray source about 7300 light years away (Webster and Murdin 1972). The existence of a supermassive black hole at the center of our galaxy was confirmed in 1974 with the discovery of the compact radio source Sagittarius A* (Balick and Brown 1974, Cox and Reynolds 2004). Around the same time, Nikolai Shakura and Rashid Sunyaev predicted that many black holes would be accompanied by thin accretionary disks, with matter spiralling inward toward the event horizon (Shakura and Sunyaev 1973). This phenomenon was observationally verified in the 1990s (Marsh et al 1994). Most astronomers now believe that supermassive black holes exist at the center of every large galaxy, and a multitude of smaller-mass black holes occur elsewhere (Kormendy and Richstone 1995).

Figure 1B: Elliptical galaxy M87 emitting an astrophysical jet (Hubble Space Telescope Image).

Figure 1C: The Circinus galaxy, a type II Seyfert galaxy (Hubble Space Telescope Image).

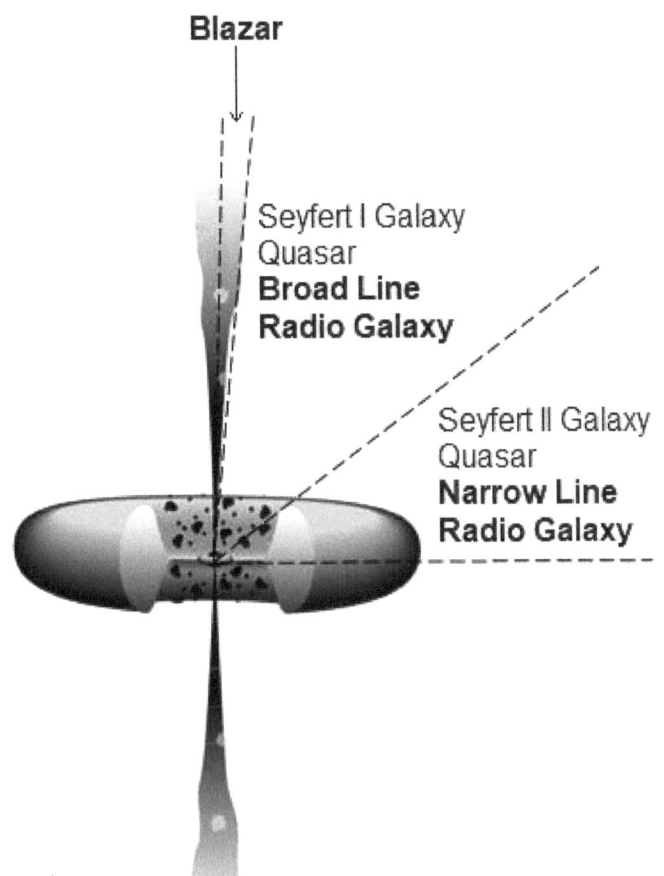

Blazar

Seyfert I Galaxy
Quasar
**Broad Line
Radio Galaxy**

Seyfert II Galaxy
Quasar
**Narrow Line
Radio Galaxy**

Figure 1D: An accretion disk with a polar jet. The classification of different AGN types is based on the viewing angle.

In addition to accretionary disks, some supermassive black holes have polar jets that emit matter at relativistic speeds. Sources of astrophysical jets are known as Active Galactic Nuclei (AGN). The first astrophysical jet was discovered in 1918 by Heber Curtis, who noticed a peculiar feature extending out of elliptical galaxy M87, shown in figure 1B. Figure 1C shows the Circinus galaxy, a type II Seyfert galaxy, where the source of the jet is thought to be obscured by a torus of dust (Maran 1991). Most modern astronomers believe that the material in astrophysical jets is mostly electron-proton plasma, the building blocks of hydrogen atoms (Sikora and Madejski 2000, Madejski and Sikora 2016). The power output of jets emerging from AGNs ranges from 10^{42} to 10^{46} erg s^{-1}, which is comparable to the luminosity of entire galaxies (Romero 2021). Although we don't really know much about what actually happens inside rapidly rotating supermassive black holes at the center of AGNs, they appear to be taking in stellar matter from their accretionary disks and accelerating it to significant fractions of the speed of light. This eventually spaghettifies the matter, allowing it to be split apart at the subatomic level (Pinochet 2022). The black holes then occasionally spit out immense bursts of energy, electrons, positrons and protons, from their polar jets. It's not exactly known why AGNs produce these jets, but some physicists have associated it with an effect of general relativity known as frame dragging (Miller-Jones et al 2019).

AGN Unification

Active Galactic Nuclei are now thought to be a diverse set of extragalactic objects that includes Quasars, Blazars, Seyfert galaxies and Radio galaxies. Unified AGN models differentiate these objects based on the orientation of their jets relative to the observer, as illustrated in Figure 1D (Antonucci 1993). Seyfert II galaxies, like the Circinus galaxy in figure 1C, represent edge-on disks with jets extending upward and/or downward from the galaxy's center, as seen by the observer. Our view of Seyfert II galaxies does not include the polar jet itself, as it is obstructed by a torus of accretionary material. On the other hand, blazars have jets that shoot directly towards the observer. The jets of Quasars and Seyfert I galaxies are inclined by a small angle. The light received from blazars and quasars tends to be extremely redshifted. Radio galaxies and Blazars emit large amounts of radio waves while Seyfert galaxies are radio quiet. AGN Unification theories are important as they suggest that astrophysical jets are more common in the universe than originally believed.

Conclusions

Before Active Galactic Nuclei were observed and understood, the abundance of light elements in the universe helped the Big Bang Theory gain support over the Steady State theory. However, it now seems that astrophysical jets produced by AGNs cause a continuous replenishment of light elements. Therefore, the observed cosmic abundances of ^{2}H, ^{3}He, He and ^{7}Li may not reflect the state of nucleosynthesis following the Big Bang but simply represent a balance between hydrogen being produced by astrophysical jets and heavier elements being produced by stars. This gives vision to a sustainable universe where black holes at the center of AGNs recycle matter and replenish the cosmic supply of light elements.

References

Antonucci,R.(1993).Unified models for active galactic nuclei and quasars.*Annual review of astronomy and astrophysics*,31(1),473-521.

Alpher,R.A.,Bethe,H.,Gamow,G.(1948).The origin of chemical elements.*Physical Review*,73(7),803.

Balick,B.,Brown,R.L.(1974).Intense sub-arcsecond structure in the galactic center.*The Astrophysical Journal*,194,265-270.

Beall,J.H.(2014).A review of astrophysical jets.*Acta Polytechnica CTU Proceedings*,1(1),259-264.

Boesgaard,A.M.,Steigman,G.(1985).Big bang nucleosynthesis-Theories and observations.*Annual Review of Astronomy and Astrophysics*,23,319-378.

Clayton,D.D.(1968).Principles of Stellar Evolution and Nucleosynthesis.*Chicago*.

Cox,D.P.,Reynolds,R.J.(2004).How Does the Galaxy Work?: A Galactic Tertulia with Don Cox and Ron Reynolds(Vol. 315).Springer Science & Business Media.

Croswell,K.(1995).The alchemy of the heavens:searching for meaning in the milky way(1st Anchor books).*Anchor Books*.

Cyburt,R.H.,Fields,B.D.,Olive,K.A.(2008).A bitter pill: the primordial lithium problem worsens.*arXiv preprint:0808. 2818*.

Fields,B.D.(2011).The primordial lithium problem.*Annual Review of Nuclear and Particle Science*,61,47-68.

Finkelstein,D.(1958).Past-future asymmetry of the gravitational field of a point particle.*Physical Review*,110(4),965.

Hoyle,F.,Tayler,R.J.(1964).The mystery of the cosmic helium abundance.*Nature*,203(4950),1108-1110.

Kenath,A.Sivaram,C.(2008).Some aspects of Primordial Nucleosynthesis.10.13140/RG.2.1.4761.9929.

Kolb,E.W.,Turner,M.S.(2018).The early universe.*CRC press*.

Kormendy,J.,Richstone,D.(1995),Inward Bound-The Search For Supermassive Black Holes In Galactic Nuclei.*Annual Review of Astronomy and Astrophysics*,33:581.

López-Corredoira,M.(2022).The abundance of light elements. In *Fundamental Ideas in Cosmology:Scientific, philosophical and sociological critical perspectives*.IOP Publishing.

Madejski,G.,Sikora,M.(2016).Gamma-ray observations of active galactic nuclei.*Annual Review of Astronomy and Astrophysics*,54,725-760.

Maran,S.P.(1991).*The Astronomy and astrophysics encyclopedia*.Van Nostrand Reinhold.

Marsh,T.R.,Robinson,E.L.,Wood,J.H.(1994).Spectroscopy of: the mass of the black hole and an image of its accretion disc.*Monthly Notices of the Royal Astronomical Society*, 266(1),137-154.

Miller-Jones,J.C.,Tetarenko,A.J.,Sivakoff,G.R.,Middleton,M.J.,Altamirano,D.,Anderson,G.E.,Tudose,V.(2019). A rapidly changing jet orientation in the stellar-mass black-hole system V404 Cygni.*Nature*,569(7756),374-377.

Montgomery,C.,Orchiston,W.,Whittingham,I.(2009).Michell, Laplace and the origin of the black hole concept.*Journal of Astronomical History and Heritage*.12(2):90–96.

Peebles,P.J.E.(1966).Primeval helium abundance and the primeval fireball.*Physical Review Letters*,16(10),410.

Pinochet,J.(2022).The little robot,black holes,and spaghettification.*Physics Education*,57(4),045008.

Reeves,H.(1974).On the origin of the light elements.*Annual review of astronomy and astrophysics*,12(1),437-470.

Romero,G.E.(2021).The content of astrophysical jets.*Astronomische Nachrichten*,342(5),727-734.

Schwarzschild,K.(1916).Uber das Gravitationsfeld eines Massenpunktes nach der Einstein'schen Theorie.Berlin. *Sitzungsberichte*,18.

Shakura,N.I.,Sunyaev,R.A.(1973).Black holes in binary systems.Observational appearance.*Astronomy and Astrophysics*,24,337-355.

Sikora,M.,Madejski,G.(2000).On pair content and variability of subparsec jets in quasars.*The Astrophysical Journal*, 534(1),109.

Spergel,D.N.,Verde,L.,Peiris,H.V.,Komatsu,E.,Nolta,M.R., Bennett,C.L.,Wright,E.L.(2003).First-year Wilkinson Microwave Anisotropy Probe(WMAP)*observations:determination of cosmological parameters.*The Astrophysical Journal Supplement Series*,148(1),175.

Wagoner,R.V.,Fowler,W.A.,Hoyle,F.(1967).On the synthesis of elements at very high temperatures.*The Astrophysical Journal*,148,3.

Webster,B.L.,Murdin,P.(1972).Cygnus X-1:a Spectroscopic Binary with a Heavy Companion?*Nature*,235(5332),37-38.

Woosley,S.E.,Arnett,W.D.,Clayton,D.D.(1973).The explosive burning of oxygen and silicon.*Astrophysical Journal Supplement*.

A possible solution to the dark matter problem in spiral galaxies

Abstract

Presented here is an alternate model for galactic evolution in which spiral galaxies evolve from elliptical galaxies in a gradual flattening process that takes hundreds of billions of years and elliptical galaxies form from mergers of large galaxies. This model allows the universe to sustain itself as spiral and elliptical galaxies are both being continuously replenished. When a main sequence star in an elliptical galaxy reaches the end of its life, its burnt-out core continues orbiting in the same path. Therefore, as elliptical galaxies slowly flatten and mature into spiral galaxies, its gas, dust and new stars settle closer to the galactic equator while large collections of Massive Compact Halo Objects (MACHOs) are left in the halo as dark matter. Although extensive microlensing surveys of the Milky Way's halo have accounted for only a small fraction of our galaxy's missing mass, it is possible that sizeable populations of old, faint, cool white dwarfs may have so far evaded detection. Using this alternate scenario of galactic evolution, the minimum age of the Milky Way is calculated to be ~32 Gyr, which is incompatible with standard Big Bang cosmology. However, if the ~14 Gyr time constraint imposed by the Big Bang Theory is removed, this alternate model of galactic evolution is applied, and the existence of a large population of old, faint white dwarfs in our galaxy's halo is considered, then a resolution to the dark matter problem in spiral galaxies may be possible.

Introduction

Knowledge of distant galaxies advanced greatly with Edwin Hubble's observations at the Mount Wilson Observatory in the 1920s. Hubble developed a classification system for distant galaxies, now known as the tuning fork diagram, which divided galaxies into four types: elliptical, lenticular, spiral and barred spiral (Hubble 1926). A fifth type, irregular galaxies, was included for galaxies that didn't fit on the tuning fork, as you can see in figure 2A. Expanded classification schemes were devised in the 1950s but the basic tuning fork shape was retained (Vaucouleurs 1959).

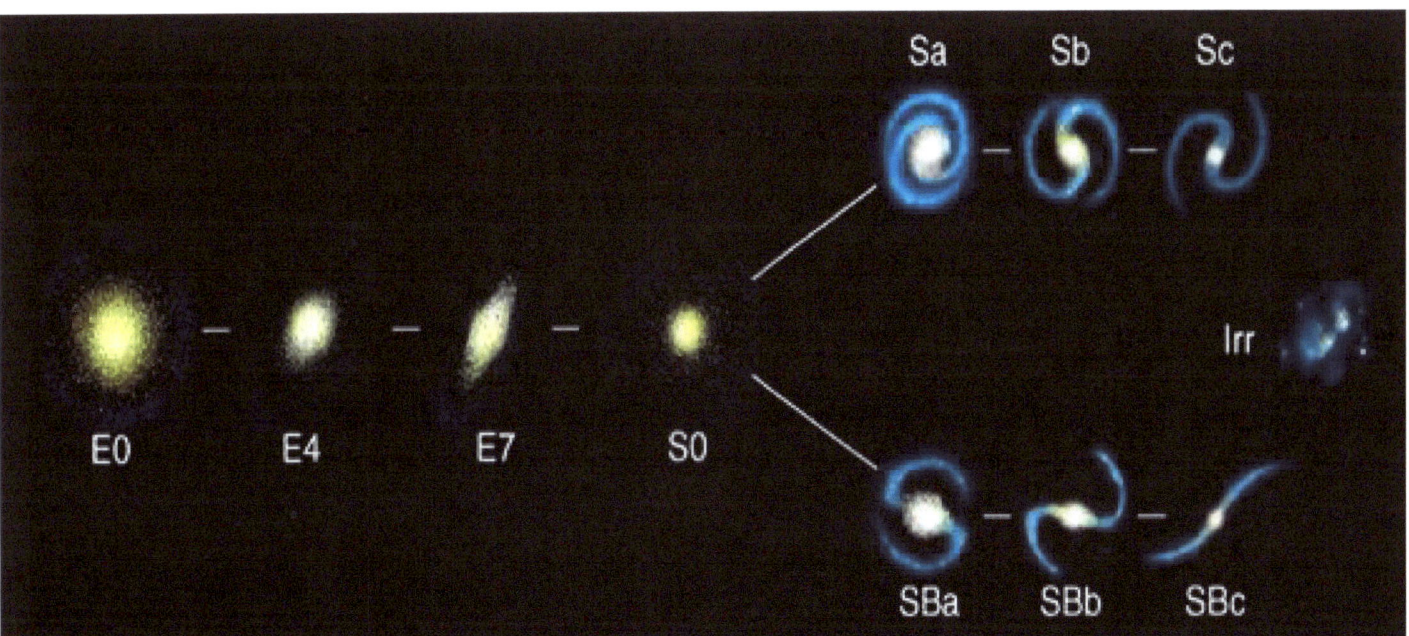

Figure 2A: Hubble's original tuning fork galaxy classification diagram (Hubble 1926).

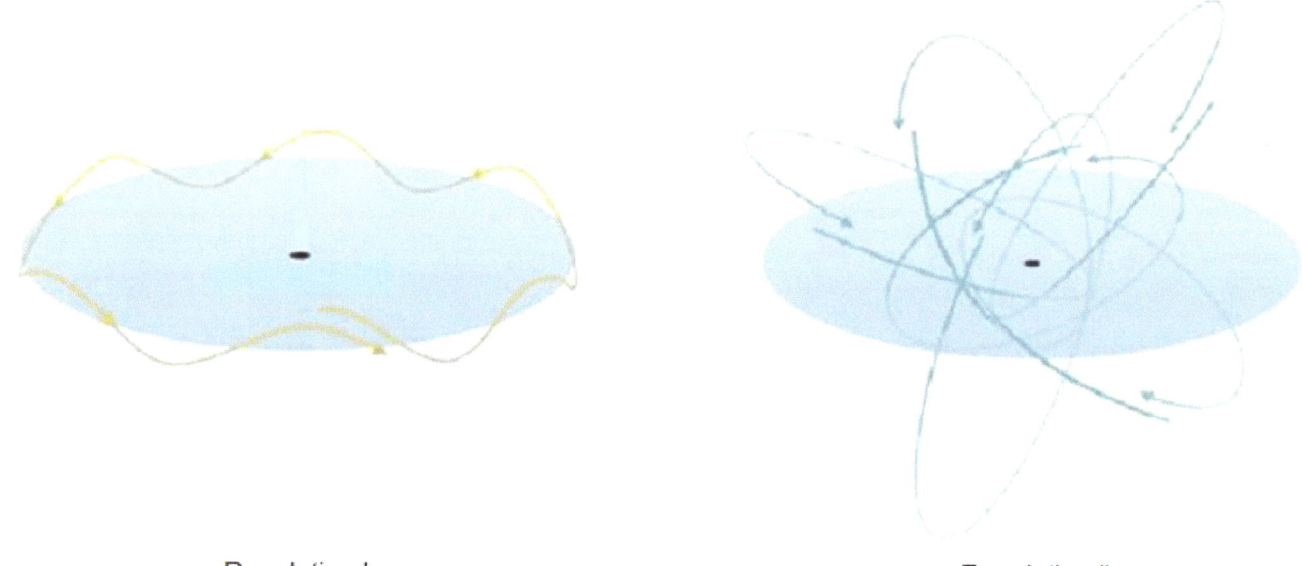

Population I   Population II

Figure 2B: Example orbits of population I and II stars in the Milky Way galaxy. Diagram is not to scale.

Hubble originally believed that elliptical galaxies would eventually transform into spiral galaxies as the mutual gravitational attraction of stars within them would cause the system to slowly condense and spin in order to preserve its total angular momentum (John 2006). Conceptually, this was the same process hypothesized for the formation of the solar system but applied on larger distance scales. This disk formation and flattening process as spiral galaxies mature may be reflected in the concept of Population I and II stars, an idea originally described by Jan Oort in 1926 but first published by Walter Baade in 1944 (Baade 1944). Baade defined these two categories of stars in the galaxy based on their metallicity and their orbital kinematics, as illustrated in figure 2B.

Population I stars like our Sun are younger than population II stars and have higher metallicities. They have low eccentricity orbits that oscillate with respect to the galactic plane somewhat like a carousel's horse. The entire group of population I stars is confined to our galaxy's thin disk, which extends about 325 parsecs on each side of the galactic plane (Schneider 2006). Population II stars reside mostly in the galaxy's thick disk, which has a scale height of around 1.5 kpc (Schneider 2006). They are, on average, an older group of stars with more disordered, eccentric and inclined orbits. Population II also includes the majority of stars in globular clusters. The two populations of stars vary greatly in metallicity, with population I averaging around z = 0.2 and population II averaging around z = 0.001 (Bartasiute et al 2003, Schneider 2006). Since

younger stars with higher metallicities have a tighter distribution around the galaxy's plane, this suggests that over time the Milky Way's luminous material is gradually flattening from one stellar generation to the next. A hypothetical group of metal-poor population III stars was discussed in the 1980s but has not been supported by observational evidence (Bond 1981).

Hubble's notion of spiral galaxies forming from the collapse of large clouds of stars was echoed by Olin Eggen, Donald Lynden-Bell and Allan Sandage in the 1960s (Eggen et al 1962). However, the cosmological debate between the Big Bang Theory and the Steady State Theory fizzled out soon after the discovery of the Cosmic Microwave Background Radiation in 1965 (Penzias and Wilson 1965). This dissuaded most astronomers from the intuitive galactic evolution models conceived by Hubble and Sandage. Most astronomers today reject the notion that spiral galaxies could evolve from elliptical galaxies as it would be impossible to contain the necessary chain of events within the ~14 Gyr timeframe allotted by the Big Bang Theory. Instead, modern-day supporters of the Big Bang Theory suggest that complex spiral galaxies such as our own formed directly from the initial conditions of the young universe (Avila-Reese et al 1998).

The existence of dark matter in the halo of disk galaxies was originally theorized by Jan Oort and his contemporary Fritz Zwicky in the 1930s (Zwicky 1933). After examining the orbits of stars in the stellar neighbourhood and the distribution of stars in other galaxies, Oort noticed that the orbital dynamics of stars in spiral galaxies didn't follow Kepler's third law (Oort 1940). This sentiment was echoed by other astronomers for a few decades and by the late 1960s, most astronomers agreed that something was not right about the rotation curves of all spiral galaxies (Babcock 1939, Volders 1959, Rubin et al 1980). As you can see in figure 2C, the orbital velocities of stars in the outer regions of many spiral galaxies are much faster than they should be according to Kepler's laws. By 1974, several astronomers had proposed that this discrepancy was due to a mysterious missing mass in the haloes of spiral galaxies, now known as dark matter (Einasto et al 1974, Ostriker et al 1974). Today's astronomers estimate that the dark matter component of the Milky Way's mass is somewhere between 64% and 95% (Watkins et al 2019, Jiao et al 2023).

Three major explanations for the identity of the mysterious missing mass have been proposed. Weakly Interacting Massive Particles (WIMPs), first written about in 1978, and axions, proposed in 1977, are hypothetical particle groups that were theorized to have collectively high mass but are so far undetectable to us (Peccei and Quinn 1977, Faulkner and Gilliland 1985). Although the WIMP theories were put together elegantly, they have suffered from a lack of observational and experimental evidence. By the late 2000s, they had been discounted by most academics (Siegel 2019).

In the early 1980s, Israeli physicist Mordehai Millgrom proposed a theory called Modified Newtonian Dynamics (MOND), which involves an alteration of physical laws at large distances (Milgrom 1983). Although supporters of MOND claim that a wide range of astrophysical phenomena are consistent with its framework, the theory requires some ad hoc additions to general relativity (Scott et al 2001). Most astronomers today discount MOND theories based on theoretical issues and a shortage of observational tests (Scott et al 2001). However, using similar lines of thought, some recent publications have claimed to be able to reconcile the dark matter problem with general relativity alone (Crosta et al 2020, Ludwig 2021).

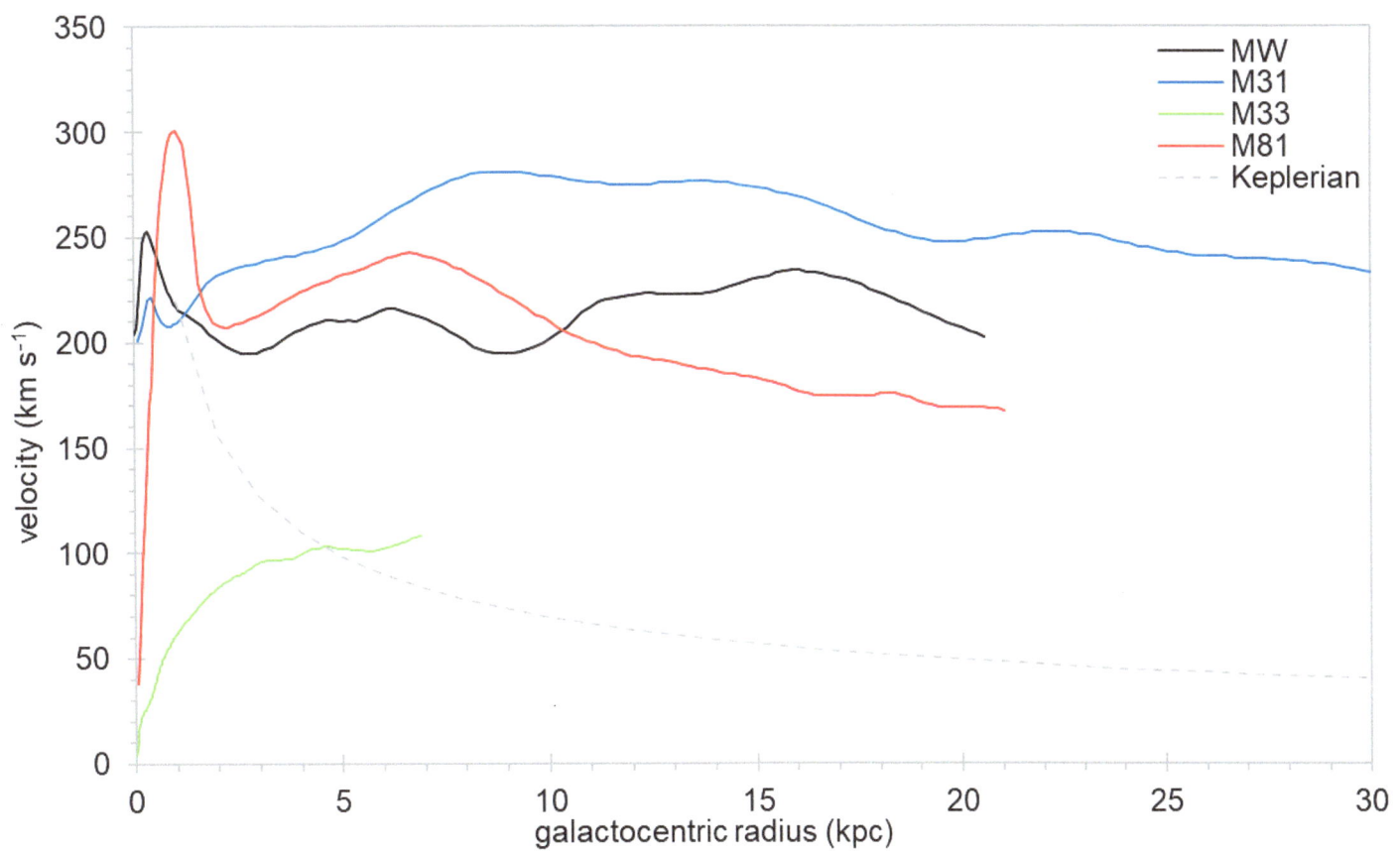

Figure 2C: Galactic rotation curves of the Milky Way and three other spiral galaxies. (Sofue 1997)

	mass (M$_\odot$)	radius (R$_\odot$)	luminosity (L$_\odot$)	lifetime (yr)
Red dwarf	0.08 - 0.45	0.1 - 0.7	3×10^{-4} - 0.08	0.8 - 12 trillion
White dwarf	0.17 - 1.33	0.008 - 0.02	1×10^{-4} - 0.0295	10 billion - 10^{15}
Brown dwarf	0.012 - 0.076	0.118 - 0.123	1×10^{-5}	∞
Neutron star	1.1 - 2.16	1.5×10^{-5} - 2.2×10^{-5}	1×10^{-9} - 1×10^{-6}	10 - 500 billion
Low mass black hole	5 - 10^5	4.3×10^{-5} - 0.01	0	10^{64} - 10^{100}
Black dwarf	0.17 - 1.33	0.008 - 0.02	0	∞

Figure 2D: Characteristics of different types of low luminosity stellar objects (MACHOs).

Massive compound halo objects (MACHOs) were first suggested as an explanation for dark matter in the 1990s (Griest 1991). MACHOs are a broad group of low-luminosity stellar objects that include red dwarfs, white dwarfs, brown dwarfs, neutron stars, low-mass black holes and the so far theoretical black dwarfs. Generally speaking, MACHOs have low luminosities and long lifetimes compared to main sequence stellar objects. A summary of the characteristics of the different classes of MACHO candidates is provided in figure 2D.

A number of high-profile extensive MACHO detection projects undertaken in the last few decades have garnered negative results. Both the MACHO and EROS halo surveys did not report any microlensing events with durations more than a few hours, leading most astronomers to conclude that objects with masses between 10^{-7} and 10^{-3} M$_\odot$ can't account for more than 10% of the halo's dark matter (Torres et al 2008). This essentially ruled out the possibility that planets and brown dwarfs could represent a significant portion of the missing mass. The majority of microlensing events observed by halo survey projects corresponded with masses in the range between 0.15 and 0.50 M$_\odot$, which overlaps much of the red dwarf and white dwarf ranges (Alcock et al 2000).

Of the three types of proposed solutions to the dark matter problem in spiral galaxies, the MACHO explanation is on the right side of Occam's Razor as it does not require any adjustments or additions to known physical laws. Therefore, it should be considered the safest explanation and this article will explore the possibility that 100% of the dark matter in the haloes of spiral galaxies exists in the form of MACHOs. As a group, they could represent the long burnt-out cores of older generations of stars that shone during earlier stages of galactic evolution. In this way, the Milky Way's halo could act as a sort of stellar graveyard filled with the remnant cores of main sequence stars that formed when our galaxy was more elliptical in form.

Formation of Spiral Galaxies

In short, the basic notion of elliptical galaxies evolving into spiral galaxies is a magnified version of the standard explanation for the formation of the solar system. Spherical collections of gas and dust collapse due to the mutual gravitational attraction of the material within them. As they do this, dense matter collects at the center and the system as a whole rotates faster and faster to conserve the total angular momentum. As elliptical galaxies contract and transform into spirals, the distribution of gas, dust and luminous matter gradually moves from an ellipsoidal arrangement into a relatively thin disk while old, burnt-out star cores remain in the halo. After long periods of time, successive generations of new stars settle into lower eccentricity and lower inclination orbits.

A star's orbit doesn't actually change to any significant amount in inclination or eccentricity during its lifetime and collisions between star systems are rare. However, a galaxy's collection of less inertial dust and gas clouds gradually shifts toward the galactic plane, the galactic center and the spiral arms to evolve a galaxy's visible structure. Almost all star formation in our galaxy occurs in molecular clouds near the spiral arms and the galactic plane, where gas and dust is most condensed. When new stars form from clouds of interstellar gas, they generally fall into orbits that are less inclined and eccentric than stars in the previous generation. We are actually only able to perceive the last few stellar generations of this multi-billion-year flattening process in differences between population I and II stars.

Also of note is that more than 75% of main sequence stars in spiral galaxies are red dwarfs, which accounts for almost 50% of the main sequence star mass in our galaxy (Ledrew 2001). Red dwarfs can burn hydrogen continuously for up to 12 trillion years, so many of them may be much older than previous thought (Adams and Laughlin 1997).

Globular Clusters

Globular clusters are dense collections of thousands of stars that tend to exist near the bulge of spiral galaxies. They are 1.8×10^3 to 1.6×10^6 M☉ in mass and 8.5×10^3 to 5.0×10^5 L☉ in luminosity (Secker 1992, Nataf et al 2019). The average age of stars inside the Milky Way's globular clusters is older than the rest of our galaxy, and at $z = -0.1$ to -2.3, they are significantly lower in metallicity (Nataf et al 2019). Stars in globular clusters are also more tightly bound to each other than they are to the rest of the galaxy, although some stars in the outer shells of globular clusters can be tidally stripped on occasion (Benacquista and Downing 2013). Some globular clusters have unusual retorgrade orbits, which suggests that they may result from the capture of satellite galaxies (Bekki and Freeman 2003). The level of virial relaxation that globular clusters are observed to have has led several astronomers in the past to estimate an average age close to 17 Gyr (Sandage 1982, Chaboyer et al 1995). Most modern astronomers dispute this as it is higher than current estimates for the age of the universe, instead claiming that globular clusters started forming 13.5 Gyr ago, not long after the Big Bang (Jimenez et al 1996).

Black Dwarfs

A significant amount of the halo's dark matter could be represented by a population of old, faint, cool white dwarfs (Oppenheimer et al 2001). This would make sense as white dwarfs in the halo are difficult to detect due to their faintness, relatively small size and high surface gravity which affects their atmospheres. As they age, white dwarfs gradually cool and become fainter and denser until they produce so little light and heat that they be classified as black dwarfs (Caplan 2020). Theoretically, black dwarfs represent the eventual endpoint of all main sequence stars between approximately 0.08 and 10 M☉, which involves the vast majority of stellar mass in the galaxy (Heger et al 2003, Nakano 2013).

Many astronomers have disputed the notion that ancient white dwarfs can account for a significant portion of the halo's dark matter (Blaineau et al 2022). Yet in the last few decades a number of very cool white dwarfs have been discovered, including some with massive invisible companion stars (Kilic et al 2009, Van Oirschot et al 2014, Darmos 2019, Pallathadka et al 2023). The unseen companions are often pinned as neutron stars or low mass black holes as black dwarfs are considered theoretically impossible because the universe is too young for any white dwarfs to have reached this stage in their evolution (Caplan 2020). However, if one ignores time constraints imposed by the Big Bang Theory, then it is plausible to suggest that a great multitude of old, faint, cool white dwarfs and invisible black dwarfs could exist in the halo of our galaxy.

Age of the Milky Way

It is a challenge to explain how a complex spiral galaxy like the Milky Way could form within a timeframe of ~14 Gyr. If we define one stellar generation to be ~10 Gyr, the lifetime of an average star like our Sun, then 14 Gyr only allows for two stellar generations at most.

The total mass of all luminous matter in our galaxy, which includes the stars, gas and dust, is ~7.5×10^{10} M☉ (Price-Whelan 2017) According to recent measurements from the GAIA spacecraft, the total mass of the Milky Way could be as high as 1.54×10^{12} M☉ and as low as 2.06×10^{11} M☉. (Watkins et al 2019, Jiao et al 2023). This places the dark matter content of the Milky Way between 64% and 95%, the vast majority of which would be concentrated in the halo.

The oldest stars in population I are ~10 Gyr old (Franknoi et al 2016). One of the oldest known stars in population II, HD140283, was originally estimated to be ~14.5 Gyr old (Bond et al 2013). To generate a crude calculation of the Milky Way's age, think of population I and II as representing the last two stellar generations: the luminous matter in the galaxy which is at most ~15 Gyr old. If we use the lower estimate for the dark matter content in the Milky Way, 64%, and assume that this is entirely attributed to MACHOs, then the halo would contain about three and a half previous stellar generations worth of burnt-out star cores. This would place the Milky Way's age at ~32 Gyr. Using 95%, the higher estimate for the dark matter content in our galaxy, the halo could contain around forty previous stellar generations. That would place the age of our galaxy at around 200 Gyr.

Figure 2E: The mice galaxies, NGC 4676, is an example of a galaxy merger in progress. (NASA/ESA Hubble Space Telescope Image)

Elliptical Galaxies

Most astronomers today accept that elliptical galaxies form from mergers of two of more progenitor galaxies, an idea first theorized by the Toomre brothers in 1972 and supported by computer simulations in the 1980s (Toomre and Toomre 1972, Farouki and Shapiro 1982, Negroponte and White 1983, Barnes 1989). Figure 2E shows a pair of galaxies nicknamed the mice, which give us a snapshot of what a galaxy merger may look like.

Gravitational interactions between merging galaxies rip disk and spiral structure apart completely and produce enormous clouds of gravitationally bound stars with orbits ranging greatly in latitude, distance and speed. As a result, star orbits in elliptical galaxies have wider ranges of eccentricity and inclination compared to spiral galaxies. Galaxy mergers may be quite violent, and the interaction between the two sets of stars, gas and dust might cause shock waves that initiate new phases of star formation (Hopkins et al 2013). Regardless of which type of galaxies are involved in a merger, the resulting elliptical galaxy typically bears no resemblance to either of its progenitors (Barnes 1989).

The dark matter issue observed in spiral galaxies is not seen to the same extent in elliptical galaxies (Binney et al 1990). As velocity-rotation curves cannot be easily obtained from elliptical galaxies, light-to-mass ratios can serve as an alternative way of estimating dark matter content in elliptical galaxies. A 2006 study of elliptical and lenticular galaxies calculated that on average they contain about 70% luminous matter by mass (Cappellari et al 2006). This contrasts greatly with the 5 to 36% luminous matter content in our own galaxy. This discrepancy seems to support the basic notion that elliptical galaxies evolve into spiral galaxies over long periods of time and as they do, their light-to-mass ratios gradually decrease.

Yet in order for this model of galactic evolution to work, there must be a process during the merger of galaxies and the formation of elliptical galaxies that is able to transform some dark matter stored in galactic haloes back into luminous matter. Without this process, the universe as a whole would eventually become filled with nothing but dark matter and there would be little discrepancy between light-to-mass ratios in spiral and elliptical galaxies. It's possible that violent clashes between two dark matter haloes involve intense tidal interactions between MACHOs that cause the release of non-luminous matter, making it available again as fuel for future stellar generations. At this point it isn't clear what such a process would look like in detail, but it seems likely that the merger of two galactic haloes would be more effective at converting non-luminous matter into star fuel if they were primarily composed of black dwarfs rather than low-mass black holes.

Conclusions

This article explored the possibility of large elliptical galaxies evolving into spiral galaxies over timescales of at least 32 Gyr. This extended timescale is sufficient for allowing large amounts of non-luminous mass in the form of MACHOs to accumulate in the haloes of spiral galaxies. An overview of population I and II stars, the existence of old white dwarfs in the halo, and a comparison of light-to-mass ratios in elliptical and spiral galaxies has provided support for this conjecture.

This alternative model of galactic evolution offers a solution to the dark matter problem in spiral galaxies without altering any established physical laws or inventing new types of hypothetical particles. It puts forward the idea that there have been more than two stellar generations in our galaxy's past and that long-burnt-out star cores from previous stellar generations could still be travelling in orbits similar to their progenitors. Although this model does not yet have abundant support, it could gain plausibility if future research reveals a higher density of red, white or black dwarf stars in the halo of our galaxy.

This alternative galactic evolutionary scenario presents a vision of a sustainable universe, with new spiral galaxies evolving from elliptical galaxies and new elliptical galaxies forming from mergers. This is in fact a rather intuitive model for galactic evolution that resembles earlier ideas proposed by Edwin Hubble and Allan Sandage. By removing the time constraint imposed by expansionist models of the universe and restoring these older ideas of galactic evolution, it may be possible to reconcile the dark matter problem in spiral galaxies.

Acknowledgments

Ian Wilckie is thanked for reviewing this article and providing insightful comments.

References

Adams,F.C.,Laughlin,G.(1997).A dying universe:the long-term fate and evolution of astrophysical objects.*Reviews of Modern Physics*,69(2),337.

Alcock,C.,Allsman,R.A.,Alves,D.R.,Axelrod,T.S.,Becker,A.C., Bennett,D.P.,Welch,D.et al.(2000).The MACHO project: microlensing results from 5.7 years of Large Magellanic Cloud observations.*The Astrophysical Journal*,542(1),281.

Avila-Reese,V.,Firmani,C.,Hernández,X.(1998).On the formation and evolution of disk galaxies: Cosmological initial conditions and the gravitational collapse.*The Astrophysical Journal*,505(1),37.

Baade,W.(1944).The Resolution of Messier 32,NGC 205, and the Central Region of the Andromeda Nebula.*The Astrophysical Journal*,100,137.

Babcock,H.W.(1939).The rotation of the Andromeda Nebula. Lick observatory bulletin,19,41-51.

Barnes,J.E.(1989).Evolution of compact groups and the formation of elliptical galaxies.*Nature*,338(6211),123-126.

Bartašiūtė,S.,Aslan,Z.,Boyle,R.P.,Kharchenko,N.V.,Ossipkov, L.P.,Sperauskas,J.(2003).Stellar populations of the Galactic disk:Metallicity distribution and kinematics.*Open Astronomy*,12(4),539-546.

Bekki,K.,Freeman,K. C.(2003).Formation of ω Centauri from an ancient nucleated dwarf galaxy in the young Galactic disc.*Monthly Notices of the Royal Astronomical Society*,346(2),L11-L15.

Benacquista,M.J.,Downing,J.M.(2013).Relativistic binaries in globular clusters. Living Reviews in Relativity,16,1-99.

Binney,J.J.,Davies,R.L.,Illingworth,G.D.(1990).Velocity mapping and models of the elliptical galaxies NGC 720, NGC 1052,and NGC 4697.*Astrophysical Journal,Part 1 (ISSN 0004-637X),vol.361*,78-97.

Blaineau,T.,Moniez,M.,Afonso,C.,Albert,J.N.,Ansari,R.,Aubourg,E.,Tisserand,P.et.al.(2022).New limits from microlensing on Galactic black holes in the mass range 10 M⊙< M<1000 M⊙.*Astronomy & Astrophysics*,664,A106.

Bond,H.E.(1981).Where is population III?*The Astrophysical Journal*,248,606-611.

Bond,H.E.,Nelan,E.P.,VandenBerg,D.A.,Schaefer,G.H., Harmer,D.(2013).HD 140283:A star in the solar neighborhood that formed shortly after the Big Bang.*The Astrophysical Journal Letters*,765(1),L12.

Caplan,M.E.(2020).Black dwarf supernova in the far future. *Monthly Notices of the Royal Astronomical Society*,497(4), 4357-4362.

Cappellari,M.,Bacon,R.,Bureau,M.,Damen,M.C.,Davies,R.L., De Zeeuw,P.T.,Van De Ven,G.(2006).The SAURON project-IV.The mass-to-light ratio,the virial mass estimator and the Fundamental Plane of elliptical and lenticular galaxies.*Monthly Notices of the Royal Astronomical Society*, 366(4),1126-1150.

Chaboyer,B.,Demarque,P.,Sarajedini,A.(1995).Globular cluster ages and the formation of the galactic Halo.*arXiv preprint astro-ph/9509063*.

Crosta,M.,Giammaria,M.,Lattanzi,M.G.,Poggio,E.(2020).On testing CDM and geometry-driven Milky Way rotation curve models with Gaia DR2.*Monthly Notices of the Royal Astronomical Society*,496(2),2107-2122.

Darmos,S.(2019).Quantum Gravity and the Role of Consciousness in Physics.ISBN:1533546339.Independently published.

Eggen,O.J.,Lynden-Bell,D.,Sandage,A.R.(1962).Evidence from the motions of old stars that the Galaxy collapsed.*The Astrophysical Journal*,136,748.

Einasto,J.,Kaasik,A.,Saar,E.(1974).Dynamic evidence on massive coronas of galaxies.*Nature*,250(5464),309-310.

Faulkner,J.,Gilliland,R.L.(1985).Weakly interacting massive particles and the solar neutrino flux.*The Astrophysical Journal*,299,994-1000.

Farouki,R.T.,Shapiro,S.L.(1982).Simulations of merging disk galaxies.*The Astrophysical Journal*,259,103-115.

Fraknoi,A.,Morrison,D.,Wolff,S.C.(2016).Astronomy.OpenStax.

Griest,K.(1991).Galactic microlensing as a method of detecting massive compact halo objects.*Astrophysical Journal, Part 1*(ISSN 0004-637X),vol.366,412-421.

Heger,A.,Fryer,C.L.,Woosley,S.E.,Langer,N.,Hartmann,D.H.(2003).How massive single stars end their life.*The Astrophysical Journal*,591(1),288.

Hopkins,P.F.,Cox,T.J.,Hernquist,L.,Narayanan,D.,Hayward,C.C.,Murray,N.(2013).Star formation in galaxy mergers with realistic models of stellar feedback and the interstellar medium.*Monthly Notices of the Royal Astronomical Society*,430(3),1901-1927.

Hubble,E.P.(1926).Extragalactic nebulae.*Astrophysical Journal*,64,321-369.

Jiao,Y.J.,Hammer,F.,Wang,H.F.,Wang,J.L.,Amram,P.,Chemin,L.,Yang,Y.B.(2023).Detection of the Keplerian decline in the Milky Way rotation curve.*Astronomy & Astrophysics*.678.A208

Jimenez,R.,Thejll,P.,JØrgensen,U.G.,MacDonald,J.,Pagel,B.(1996).Ages of globular clusters:a new approach.*Monthly Notices of the Royal Astronomical Society*,282(3),926-942.

John,D.(2006).Astronomy:The definitive guide to the universe.*Parragon Publishing*.

Kilic,M.,Brown,W.R.,Prieto,C.A.,Swift,B.,Kenyon,S.J.,Liebert,J.,Agüeros,M.A.(2009).The Runaway White Dwarf LP400−22 has a companion.*The Astrophysical Journal*,695(1),L92.

Ledrew,G.(2001).The real starry sky.*Journal of the Royal Astronomical Society of Canada*,95,32.

Ludwig,G.O.(2021).Galactic rotation curve and dark matter according to gravitomagnetism.*The European Physical Journal C*,81(2),1-25.

Milgrom,M.(1983).A modification of the Newtonian dynamics-Implications for galaxies.*The Astrophysical Journal*,270,371-383.

Naab,T.,Johansson,P.H.,Ostriker,J.P.(2009).Minor mergers and the size evolution of elliptical galaxies.*The Astrophysical Journal*,699(2),L178.

Nakano,T.(2013).Pre-main Sequence Evolution and the Hydrogen-Burning Minimum Mass.*50 Years of Brown Dwarfs:From Prediction to Discovery to Forefront of Research*,5-17.

Nataf,D.M.,Wyse,R.F.,Schiavon,R.P.,Ting,Y.S.,Minniti,D.,Cohen,R.E.,Frinchaboy,P.M.(2019).The Relationship between Globular Cluster Mass,Metallicity,and Light-element Abundance Variations.*The Astronomical Journal*,158(1),14.

Negroponte,J.,White,S.D.(1983).Simulations of mergers between disc–halo galaxies.*Monthly Notices of the Royal Astronomical Society*,205(4),1009-1029.

Oort,J.H.(1940).Some problems concerning the structure and dynamics of the galactic system and the elliptical nebulae NGC 3115 and 4494.*Astrophysical Journal*,91-273.

Oppenheimer,B.R.,Hambly,N.C.,Digby,A.P.,Hodgkin,S.T.,Saumon,D.(2001).Direct detection of galactic halo dark matter.*Science*,292(5517),698-702.

Ostriker,J.P.,Peebles,P.J.E.Yahil,A.(1974)The size and mass of galaxies, and the mass of the universe.*Astrophysical Journal*,193,L1.

Pearson,W.J.,Wang,L.,Alpaslan,M.,Baldry,I.,Bilicki,M.,Brown,M.J.I.,van Der Tak,F.F.S.(2019).Effect of galaxy mergers on star-formation rates.*Astronomy & Astrophysics*,631,A51.

Peccei,R.D.,Quinn,H.R.(1977).CP conservation in the presence of pseudoparticles.*Physical Review Letters*,38(25),1440.

Penzias,A.A.,Wilson,R.W.(1965).Measurement of the Flux Density of CAS a at 4080 Mc/s.*The Astrophysical Journal*,142,1149.

Price-Whelan,A.M.(2017).Gala:A Python package for galactic dynamics.*Journal of Open Source Software*,2(18),388.

Rubin,V.C.,FordJr,W.K.,Thonnard,N.(1980).Rotational properties of 21 SC galaxies with a large range of luminosities and radii, from NGC 4605/R=4kpc/to UGC 2885/R=122 kpc.*The Astrophysical Journal*,238,471-487.

Sandage,A.(1982).The Oosterhoff period groups and the age of globular clusters.III-The age of the globular cluster system.*Astrophysical Journal*,Part1,vol.252,553-581.

Schneider,P.(2006).Extragalactic astronomy and cosmology:an introduction(Vol. 146).*Berlin:Springer*.

Scott,D.,White,M.,Cohn,J.D.,Pierpaoli,E.(2001).Cosmological difficulties with modified Newtonian dynamics(or:La Fin du MOND?).*arXiv preprint astro-ph/0104435*.

Secker,J.(1992).A statistical investigation into the shape of the globular cluster luminosity distribution.*The Astronomical Journal*,104,1472-1481.

Siegel,E.(2019)The 'WIMP Miracle' hope for dark matter is dead.*Forbes magazine online*.

Sofue,Y.(1997).Nuclear-to-Outer Rotation Curves of Galaxies in the CO and HI lines.*Publications of the Astronomical Society of Japan*,49(1),17-46.

Springel,V.,Di Matteo,T.,Hernquist,L.(2005).Black holes in galaxy mergers: the formation of red elliptical galaxies.*The Astrophysical Journal*,620(2),L79.

Toomre,A.,Toomre,J.(1972).Galactic bridges and tails.*The Astrophysical Journal*,178,623-666.

Torres,S.,Camacho,J.,Isern,J.,García-Berro,E.(2008).The contribution of red dwarfs and white dwarfs to the halo dark matter.*Astronomy & Astrophysics*,486(2),427-435.

Van.Oirschot,P.,Nelemans,G.,Toonen,S.,Pols,O.,Brown,A.G.,Helmi,A.,Zwart,S.P.(2014).Binary white dwarfs in the halo of the Milky Way.*Astronomy & Astrophysics*,569,A42.

Vaucouleurs,G.D.(1959).Classification and morphology of external galaxies.In Astrophysik iv:Sternsysteme/astrophysics iv:Stellar systems (275-310).Springer,Berlin,Heidelberg.

Volders,L.M.J.S.(1959).Neutral hydrogen in M 33 and M 101. *Bulletin of the Astronomical Institutes of the Netherlands*, 14,323.

Watkins,L.L.,Van Der Marel,R.P.,Sohn,S.T.,Evans,N.W. (2019).Evidence for an Intermediate-mass Milky Way from Gaia DRs Halo Globular Cluster Motions.*The Astrophysical Journal*,873(2).

Zwicky,F.(1933).The redshift of extragalactic nebulae.*Helvetica Physica Acta*,6,110-127.

Could Hubble's law be based on extragalactic orbital motion?

Abstract

For almost a century, Hubble's law has defined humanity's understanding of the large-scale structure and dynamics of our universe. It has led most astronomers to conclude that the universe as a whole is expanding and that it began ~14 Gyr ago with an event known as the Big Bang. This article presents an alternate interpretation of Hubble's law based on the relative motion of our galaxy in a larger scale orbital system, where all the galaxies in our observable corner of the universe are orbiting a distant attractor in the general direction of the Shapley Supercluster (l = 306°, b = +30°). Evidence for this is discerned from extragalactic redshift surveys, which indicate that a collection of low redshift galaxies ($z < 0.004$) form a shape resembling a great circle centered at approximately l = 50°, b = +10°. This great circle is interpreted as an indication of a sort of extragalactic local standard of rest (XLSR) that approximates the direction of the Milky Way's velocity vector within the extragalactic orbital system. Other anisotropies among low redshift galaxies, such as an arc-shaped feature extending from l = 315°, b = +30° to l = 245°, b = -30°, may indicate anomalistic motion of our galaxy within the extragalactic orbital system. The primary ±0.00335°K anisotropy of the Cosmic Microwave Background in the general direction of the Shapley Supercluster may indicate an intrinsic energy gradient in our part of the universe rather than a direction of motion. Here, two scenarios regarding the nature of orbits in such an extragalactic system are explored and analogies are provided to aid in concept visualization. Characteristics for such an extragalactic system are extrapolated by comparing it to known characteristics of lower-level orbital systems. The alternate interpretation of Hubble's law presented here has advantages over the standard Big Bang Theory because it only relies on established physical laws and not on hypothetical concepts such as cosmic inflation and dark energy. It also distinguishes itself from Tired Light models as it does not dispute that cosmological redshift is primarily derived from Doppler recession.

Introduction

The tendency for distant galaxies to have redshifted spectra was first noticed by Vesto Slipher and Georges Lemaitre, but it was Edwin Hubble's publications in the 1920s that provided more convincing evidence for a direct relationship between the recession velocities of distant galaxies and their distance from us, now known as Hubble's law (Slipher 1913, Lemaitre 1927, Hubble 1929). This relation led most astronomers at the time to conclude that the universe as a whole expanded from a single event nicknamed the Big Bang (Eddington 1933). The development of new methods for estimating extragalactic distances, such as using Type 1a supernovae, has expanded the cosmic distance scale and confirmed that Hubble's law applies to the great majority of known distant galaxies, as shown in figure 3A on the next page.

An alternative explanation for the cosmological redshift commonly known as Tired Light has been proposed by several different authors over the last century starting with Fritz Zwicky in the late 1930s (Zwicky 1939, Marmet and Reber 1989, Cheng 2021). Tired Light hypotheses have repeatedly suffered from failed predictions and the plain fact that images of highly redshifted galaxies are not significantly blurred.

A series of papers in the late 1960s explored the possibility of the cosmological redshift being primarily based on gravitational redshift, a consequence of general relativity (Burbidge 1967, Hoyle and Fowler 1967, Zapolsky 1968). The discovery soon after that the fuzzy edges of distant quasars as well as their satellite galaxies had essentially the same redshift as the center of the quasar suggested that the bulk of the redshift is based on Doppler recession (Gunn 1971, Kristian 1973).

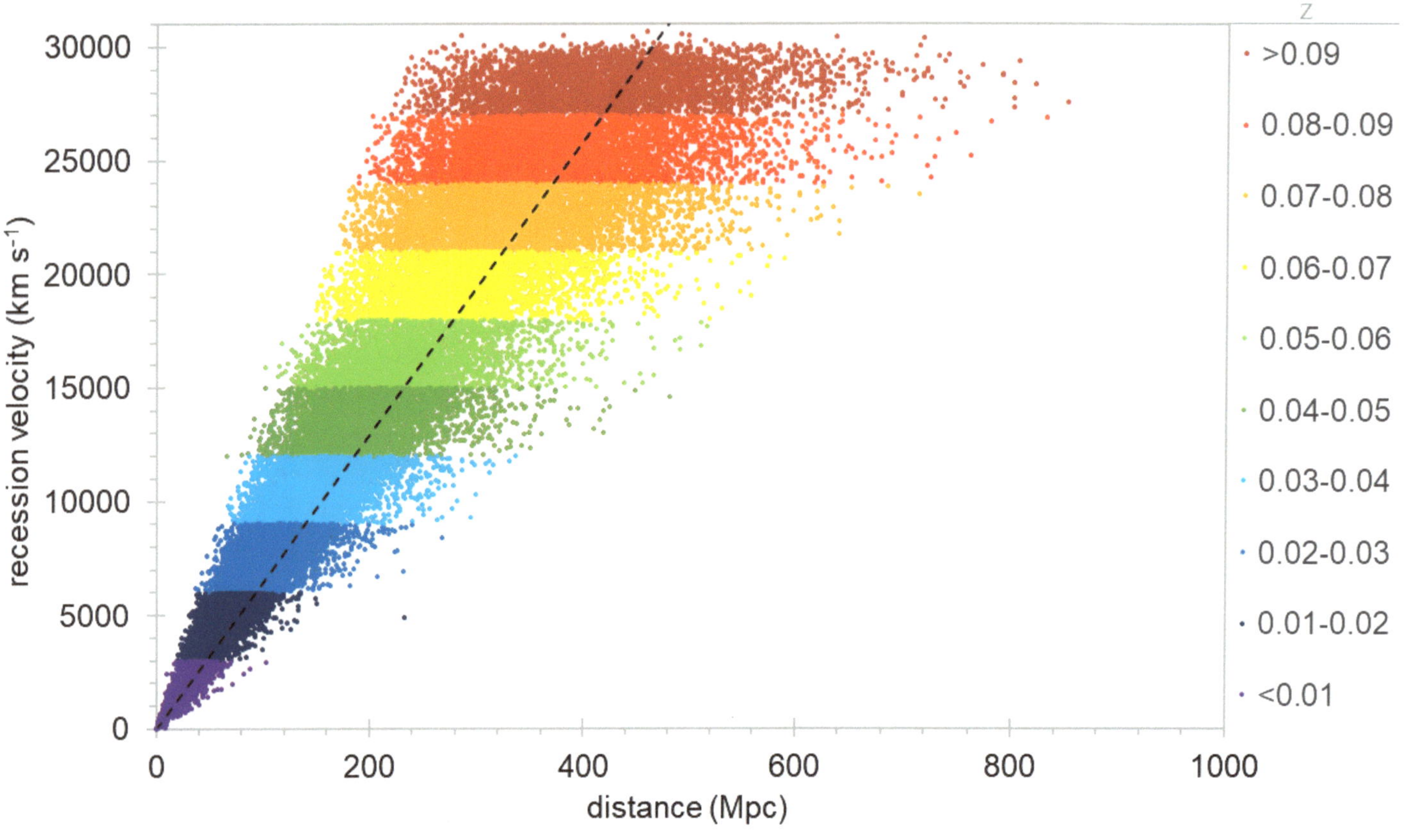

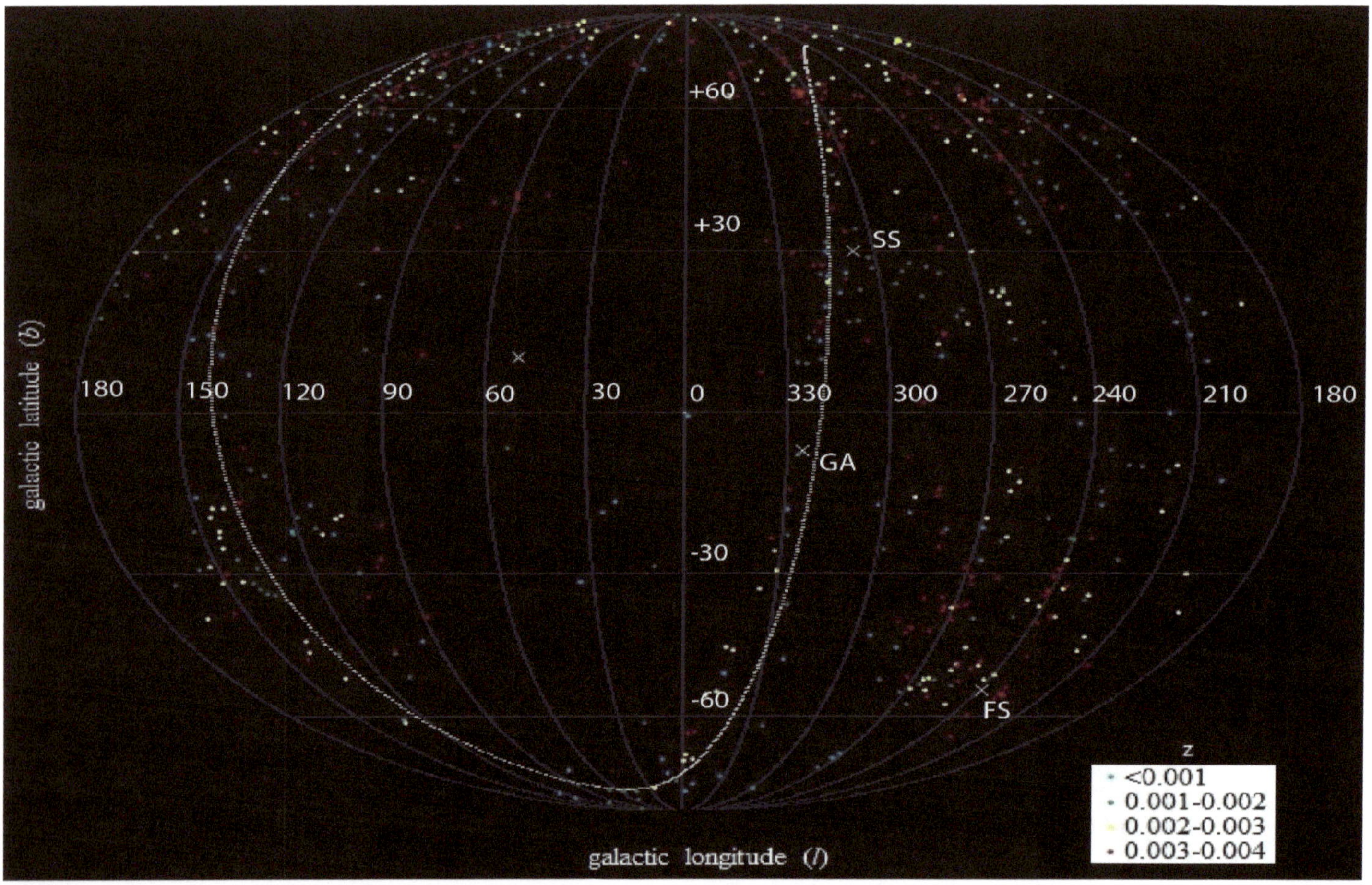

Figure 3B: Distribution of low redshift galaxies (z < 0.004) obtained from Brent Tully's Extragalactic Distance Database (Tully et al 2009). An approximate great circle shape centered around *l* = 50°, *b* = +10° is discernable. SS = Shapley Supercluster, GA = Great Attractor, FS = Fornax Supercluster. The colours in this figure are not parallels of those in figure 3A.

← **Figure 3A**: Velocity-distance data for more than 38000 galaxies with known non-redshift distance measurements. Data obtained from Brent Tully's Extragalactic Distance Database (Tully et al 2009). Here the regression line has a slope of 65 km s^{-1} Mpc^{-1}.

Although the most common interpretation of Hubble's law, the Big Bang Theory, has survived a number of challenges and a large degree of scrutiny over the last 100 years, the presentation of multiple alternate interpretations by high-profile astronomers such as Fritz Zwicky and Fred Hoyle goes to show that a significant number of astronomers have not been completely satisfied with the standard interpretation of Hubble's law.

It should also be noted that two different values for the Hubble constant have been calculated using two different methods. In 1994, Adam Reiss' group calculated a value of ~67 km s^{-1} Mpc^{-1} based on an analysis of Type 1a Supernova light curves (Riess et al 1994). In 2006, the Chandra X-ray Observatory analyzed galaxy clusters to calculate a value of ~77 km s^{-1} Mpc^{-1} (Bonamente et al 2006). All other recent estimates of the Hubble constant have essentially fallen somewhere between these two values. This discrepancy has caused a recent astronomical controversy sometimes referred to as the Hubble tension (Di Valentino et al 2021).

The following pages contain an alternative interpretation for Hubble's law not previously explored in scientific literature. It involves large-scale motion of our galaxy and all of its neighbours in orbits around a distant attractor. This interpretation of Hubble's law is offered as an alternative to the Big Bang Theory and Tired Light and may be consistent with some fractal cosmologies.

Data

Detailed redshift surveys, such as the 2MASS study, show a few distinct anisotropies in the recession velocities of galaxies in the night sky. The most important anisotropies can be inferred from the distribution of the lowest redshift galaxies (z < 0.004), which is displayed in figure 3B. A prominent feature in figure 3B is an approximate great circle shape centered at approximately $l = 50°$, $b = +10°$. Three additional anisotropies can also be inferred from the distribution of low-redshift galaxies. One of these resembles an arc-shaped streak beginning at approximately $l = 315°$, $b = +30°$ and arcing southeastward to a terminus near $l = 245°$, $b = -30°$. A patch of low redshift galaxies exists in the approximate direction of the Fornax supercluster at approximately $l = 240°$, $b = -50°$. Another patch of low redshift galaxies appears to be centered near $l = 210°$, $b = +50°$.

Although the Cosmic Microwave Background (CMB) spectrum may be the closest thing in the universe to that of a perfect blackbody, it contains a ±0.00335°K dipole anisotropy with a warm pole in the direction of Leo and a cold pole in the direction of Aquarius (Smoot 2007, Durrer 2020). After a series of astronomical tests in the 1970s, this anisotropy was attributed to the relative motion of the solar system with respect to the CMB (Rubin 1976). By the 1980s, many astronomers began to describe a scenario where our whole extragalactic neighbourhood was moving towards a mysterious center of mass ~63 Mpc away nicknamed the Great Attractor (Bertschinger and Juszkiewicz 1988, Dressler 1988). The Great Attractor's position behind the plane of our own galaxy presented difficulties for observing it directly.

Further studies of the CMB's primary anisotropy have suggested that its actual direction is toward the Shapley Supercluster, a dense collection of large, old galaxies ~200 Mpc away (Hoffman et al 2017). By subtracting the Earth's orbital velocity, the Sun's velocity within the Milky Way, and the motion of our galaxy within the local group, the velocity of our galactic neighbourhood with respect to the CMB has been estimated to be 631 km s^{-1} in the approximate direction $l = 306°$, $b = +30°$ (Hoffman et al 2017). It is possible, however, that the CMB's primary anisotropy could be based on an intrinsic temperature gradient within our part of the universe, rather than being solely based on relative kinematic motion (Abdalla 2022). After subtracting our motion within the solar system, the Milky Way and the local group of galaxies, we may simply be seeing a background energy gradient.

Interpretation

The Milky Way and all of the other galaxies in our neighbourhood could have kinematic motion, or even orbital motion, around the Shapley Supercluster or some other distant attractor. In such an extragalactic system, our own galaxy could be comoving with a set of neighbouring galaxies in a way similar to the Sun's local standard of rest. If this were the case, we could expect to see a ring of comoving galaxies spanning 360° around the night sky.

Imagine a microscopic observer riding a falling raindrop. The observer would see all raindrops below them moving away at faster and faster rates with increasing distance. All raindrops above them would also appear to be moving away as the observer raindrop accelerates downward ahead of the higher raindrops. Yet there would be a set of equialtitudinal raindrops in a ring around the observer that would not appear to be moving away. Note that in this analogy, no raindrops would appear to be approaching the observer. The great majority of them would be moving away, and with higher and higher speeds as their distances increase.

The approximate great circle in figure 3B may represent a feature analogous to the observer raindrop's equialtitudinal ring of non-receding neighbours, a sort of extragalactic standard of rest (XLSR). The great circle's center ($l = 50°$, $b = +10°$) could respresent the Milky Way's approximate velocity vector with respect to the extragalactic orbital system. It is separated from the Shapley concentration by about a quarter of the night sky, meaning that the XLSR would be moving in a direction approximately tangential to its attractor, like in all other stable gravitational orbits.

The arc-shaped feature in figure 3B arcing southeastward from $l = 315°$, $b = +30°$ to $l = 245°$, $b = -30°$ could represent a sort of angular precession of our galaxy as it slowly turns toward the Shapley attractor in its extragalactic orbit. In other words, we could be gaining ground on galaxies that are in larger orbits around our common attractor as a runner in an inside lane of a curved track gains ground on runners in the outer lanes. Another possible interpretation for this arc of low redshift galaxies is that it could represent a disk-like structural feature in the extragalactic system. In that case, observers in such an extragalactic orbital system might see lower redshifts coming from galaxies in the hypothetical extragalactic disk as their direction is slowly changing.

Two other low redshift patches in figure 3B are seen near $l = 240°$, $b = -50°$ and $l = 210°$, $b = +50°$. The first of these is near the direction of the Fornax Supercluster, which is ~20 Mpc away. The other patch doesn't seem to be associated with any noticeable features. These two low-redshift patches could be collections of galaxies that have large peculiar velocities relative to the XLSR. They may represent neighbourhoods of galaxies that are coming towards us for some unknown reason.

If the kinematic motion of our neighbourhood of galaxies is reflected in redshift anisotropies, then an alternative explanation for the CMB's dipole anisotropy is required. The truth is that there is no a priori reason to automatically attribute the CMB dipole anisotropy to the motion of our galaxy within the wider universe. The CMB dipole may simply be a minor temperature gradient within the universe unrelated to our galaxy's motion, making the side of the sky containing the Shapley Supercluster and the Great Attractor thousandths of a degree warmer compared to the opposite side. Perhaps it's even possible that instead of our galaxy moving with respect to the CMB, the CMB could be moving with respect to us.

The Hubble Flow

The alternate interpretation of Hubble's law presented here proposes that the galaxies in our neighbourhood are all moving in the same general direction with similar speeds. However, if we were able to zoom out, we might realize that what we can see with our telescopes is just a small portion of a larger-scale extragalactic system, where all visible galaxies are gradually spiralling inward and picking up speed as they edge closer to an attractor. Based on this premise, there are two scenarios that might contextually explain the redshift and CMB anisotropies. The first is based on the idea of our galactic neighbourhood moving in precessing elliptical orbits around an attractor. The second involves seeing our galactic neighbourhood as part of some sort of disperse, macroscopic accretionary disk process that covers the entirety of our part of the observable universe.

In the first scenario, we would have to be moving from apastron to periastron in our elliptical orbit around the attractor. Our galactic neighbourhood would have to be on that side of the orbit because if it wasn't approaching its periastron, then the galaxies around us would tend to be blueshifted as our orbital velocity would be faster than the further outgoing galaxies and slower than more central outgoing galaxies. For the first scenario to be true, we should consider that if we were moving towards the Shapley attractor, we might expect the central part of the Shapley concentration to be blueshifted. Since this does not appear to be the case, the only explanation would be that the actual attractor lies even further beyond the Shapley Supercluster.

In the second scenario, the behaviour of galaxies would mimic the material in accretionary disks around supermassive black holes, slowly spiraling inward towards a distant event horizon. A better analogy for such an ultra-macroscopic accretionary disk process than falling raindrops would be dirt particles circling down a steep drain. As our extragalactic neighbourhood would be getting pulled in the direction of the attractor, it would become increasingly separated in a tidal like manner due to the collective gradual motion into a deeper part of the extragalactic gravitational potential. If you think of our galaxy and its neighbours gradually circling down this hypothetical drain towards an event horizon, they would be separating in all directions around us as the whole galactic neighbourhood gradually moves into a deeper part of the steepening extragalactic gravitational potential. In this case, even the XLSR would appear to be separating, although to a much smaller extent than the rest of the galaxies in the sky.

The second scenario has an advantage over the first as it does not require us to be in a particular part of an extragalactic orbit. Yet this scenario still suggests that there should be some amount of blueshift in the central part of the Shapley concentration. Could this in fact be the reason for the ±0.00335°K primary anisotropy of the CMB?

The Next Level of the Universe

To attract the vast majority of galaxies in the observable part of our universe at such extraordinary distances, the attractor would have to be an absolutely enormous collection of mass. The motion of galaxies with respect to this mysterious attractor may simply represent the next level of gravitationally-based orbital systems in our universe. Planets orbit stars, stars orbit the center of galaxies, and galaxies orbit whatever's in the center of this distant attractor. We can extrapolate some characteristics of the Milky Way's orbit around this mysterious attractor by examining patterns in the progression of the Moon's orbit around the Earth, the Earth's orbit around the Sun, and the Sun's orbit around the galactic center, as a sort of natural pattern of gravitational systems at larger and larger levels of structure.

The Sun holds approximately 99.8% of the mass of the entire solar system and its eight main planets have non-precessing Keplerian elliptical orbits (Upgren 2013). Planetary orbits have low orbital eccentricities and inclinations, with Mercury having the highest values in the solar system for both eccentricity and inclination at 0.21 and 6° respectively (Stevenson 2004).

While the solar system has eight planets and a countless number of smaller bodies, there are approximately 500 billion stars orbiting the center of our galaxy in Rosette-shaped precessing ellipses (Pagotto et al 2021). Compared to planetary orbits, star orbits within galaxies have additional parameters, concepts and features such as the local standard of rest and vertical oscillation with respect to the galactic plane. This makes the Milky Way more like a system of objects moving in a gravitational potential as opposed to the solar system's more astrolabic layout. Yet the supermassive black hole that constitutes the Milky Way's galactic center only accounts for about 3.5 millionths of the mass of the whole galaxy (Gillessen et al 2009, Carlesi et al 2022). Star orbits in the Milky Way typically have eccentricities less than 0.5, with an average of around 0.1 (Altmann et al 2004, Mackereth and Bovy 2018). Most star orbits in our galaxy have orbital inclinations less than 0.4, but there is still a large number of stars that have more highly inclined orbits as well as a sizeable collection of dark matter in the halo of our galaxy (Altmann et al 2004).

Orbits of galaxies around the extragalactic attractor would likely have additional aspects that aren't known about yet. Telescopic observations indicate that the universe's wider structure may include flocculent contorting filaments of galaxies (Bharadwaj et al 2004). This could allow some very dynamic and complex orbital paths to exist among galaxies in the wider universe.

	number of orbiting bodies	radius (Mpc)	mass ($M_\odot$)	ratio of mass in center	eccentricity	inclination	local velocity in system (km s^{-1})	period (y)
Earth-Moon system	1	1.2×10^{-14}	3.0×10^{-6}	0.988	0.055	5.1°	3.7	0.075
Solar System	8	1.5×10^{-10}	1.00	0.998	≤ 0.21	$\leq 6.3°$	30	1
Milky Way galaxy	5.0×10^{11}	2.7×10^{-2}	1.2×10^{12}	3.5×10^{-6}	≤ 0.5	$\leq 36°$	220	2.4×10^{8}
Hubble flow	2.4×10^{32}	7.6×10^{10}	4.5×10^{30}	4.3×10^{-17}	≤ 0.79	all	1500	1.0×10^{24}

Figure 3C: Characteristics of hierarchical levels of gravitationally bound orbital systems.

Figure 3C shows a summary of the characteristics for the three levels of gravitationally-based orbital systems that we know. It shows extrapolated values for a hypothetical extragalactic orbital system by fitting a polynomial to the logarithms of the values measured from the three smaller levels and using the next value in an integer sequence. Based on these patterns, the piece of the universe that we orbit within, could contain approximately 2.4×10^{32} galaxies over truly enormous distances on the order of billions of megaparsecs. The central body of the attractor might only constitute 43 quintillionths of the mass of the entire system, yet it might weigh in at approximately 190 trillion $M_\odot$, making it more than a thousand times as massive as the largest black holes identified so far (Brockamp et al 2016, Dullo et al 2021).

Conclusions

An anisotropy among low-redshift (z < 0.004) galaxies was identified in figure 3B as a great circle shape centered near $l = 50°$, $b = +10°$. The center of this great circle has been interpreted as the Milky Way's approximate velocity vector in an extragalactic orbit around a distant attractor in the general direction of the Shapley Supercluster. The great circle itself has been interpreted as a sort of extragalactic standard of rest. Other anisotropies among low-redshift galaxies, such as an arc shape running from approximately $l = 315°$, $b = +30°$ to $l = 245°$, $b = -30°$ might relate to anomalistic motion of our galaxy within the extragalactic system. In this alternative interpretation, the ±0.00335°K dipole anisotropy of the CMB represents an intrinsic energy gradient in our piece of the universe rather than a kinematic motion of our local galactic neighbourhood.

This alternate interpretation of Hubble's law has provided an explanation for the cosmological redshift that does not require an expanding universe. The majority of distant galaxies can still be receding from us in all directions based on large-scale orbital motion around a distant attractor rather than isotropic expansion of space around a single point. This cosmological model has advantages over the Big Bang Theory as it relies only on established physical laws and not hypothetical concepts such as cosmic inflation and dark energy. It also has an advantage over Tired Light models as it does not dispute that Doppler recession is the primary reason for the cosmological redshift. That all being said, it may be compatible with some fractal cosmologies.

References

Abdalla,E.,Abellán,G.F.,Aboubrahim,A.,Agnello,A.,Akarsu,Ö., Akrami,Y.,Alestas,G.,Aloni,D.,Amendola,L.,Anchordoqui, L.A.,Anderson,R.I.et al.(2022).Cosmology intertwined:A review of the particle physics, astrophysics, and cosmology associated with the cosmological tensions and anomalies.*Journal of High Energy Astrophysics*,34,49-211.

Altmann,M.,Edelmann,H.,De Boer,K.S.(2004).Studying the populations of our Galaxy using the kinematics of sdB stars.*Astronomy&Astrophysics*,414(1),181-201.

Bertschinger,E.,Juszkiewicz,R.(1988).Searching for the great attractor.*Astrophysical Journal*,Part 2-Letters(ISSN 0004-637X),vol.334,L59-L62.

Bharadwaj,S.,Bhavsar,S.P.,Sheth,J.V.(2004).The size of the longest filaments in the universe.*The Astrophysical Journal*,606(1),25.

Bonamente,M.,Joy,M.K.,LaRoque,S.J.,Carlstrom,J.E.,Reese, E.D.,Dawson,K.S.(2006).Determination of the cosmic distance scale from Sunyaev-Zel'dovich effect and Chandra X-ray measurements of high-redshift galaxy clusters.*The Astrophysical Journal*,647(1),25.

Brockamp,M.,Baumgardt,H.,Britzen,S.,Zensus,A.(2016).Unveiling Gargantua:A new search strategy for the most massive central cluster black holes.*Astronomy & Astrophysics*,585,A153.

Burbidge,E.M.(1967).Quasi-stellar objects.*Annual Review of Astronomy and Astrophysics*,5(1),399-452.

Carlesi,E.,Hoffman,Y.,Libeskind,N.I.(2022).Estimation of the masses in the local group by gradient boosted decision trees.*Monthly Notices of the Royal Astronomical Society*, 513(2),2385-2393.

Cheng,G.(2021).The expansion of the Universe may be an illusion created by Compton scattering of free electrons.

Di Valentino,E.,Mena,O.,Pan,S.,Visinelli,L.,Yang,W.,Melchiorri,A.,Mota,D.F.,Riess,A.G.,Silk,J.(2021).In the realm of the Hubble tension-a review of solutions.*Classical and Quantum Gravity*,38(15),153001.

Dressler,A.(1988).The supergalactic plane redshift survey-A candidate for the great attractor.*The Astrophysical Journal*,329,519-526.

Dullo,B.T.,Paz,A.G.,Knapen,J.H.(2021).Ultramassive Black Holes in the Most Massive Galaxies:M BH–σ versus M BH–R b.*The Astrophysical Journal*,908(2),134.

Durrer,R.(2020).The cosmic microwave background.*Cambridge University Press*.

Eddington,A.S.(1933).The expanding universe.*Nature*,132 (3332),406-407.

Gillessen,S.,Eisenhauer,F.,Trippe,S.,Alexander,T.,Genzel,R., Martins,F.,Ott,T.(2009).Monitoring stellar orbits around the Massive Black Hole in the Galactic Center.*The Astrophysical Journal*,692(2),1075.

Gunn,J.E.(1971).On the Distances of the Quasi-Stellar Objects.*Astrophysical Journal*,vol.164,L113.

Hoffman,Y.,Pomarède,D.,Tully,R.B.,Courtois,H.M.(2017). The dipole repeller.*Nature Astronomy*,1(2),0036.

Hoyle,F.,Fowler,W.A.(1967).Gravitational red-shifts in quasi-stellar objects.*Nature*,213(5074),373-374.

Hubble,E.(1929).A relation between distance and radial velocity among extra-galactic nebulae.*Proceedings of the national academy of sciences*,15(3),168-173.

Kristian,J.(1973).Quasars as events in the nuclei of galaxies:The evidence from direct photographs.*Astrophysical Journal*,vol.179,L61.

Lemaître,G.(1927).Un Univers homogène de masse constante et de rayon croissant rendant compte de la vitesse radiale des nébuleuses extra-galactiques.*Annales de la Société Scientifique de Bruxelles*,A47,49-59.

Mackereth,J.T.,Bovy,J.,(2018).Fast estimation of orbital parameters in Milky Way-like potentials.*Publications of the Astronomical Society of the Pacific*,130(993),114501.

Marmet,P.,Reber,G.,(1989).Cosmic matter and the nonexpanding universe.IEEE Transactions on Plasma Science, 17(2),264-269.

Pagotto,I.,Krajnović,D.,denBrok,M.,Emsellem,E.,Brinchmann, J.,Weilbacher,P.M.,Kollatschny,W.,Steinmetz,M. (2021).Optical emission lines in the most massive galaxies:Morphology, kinematics, and ionisation properties.*Astronomy&Astrophysics*,649,A63.

Riess,A.G.,Press,W.H.,Kirshner,R.P.(1994).Using SN Ia light curve shapes to measure the Hubble constant.*arXiv preprint astro-ph/9410054*.

Rubin,V.C.,FordJr,W.K.,Thonnard,N.,Roberts,M.S.,Graham,J.A.(1976).Motion of the Galaxy and the local group determined from the velocity anisotropy of distant SC I galaxies.I-The data.*The Astronomical Journal*,81,687-718.

Slipher,V.M.(1913).The radial velocity of the Andromeda Nebula.*Lowell Observatory Bulletin*,vol.2,no.8,56-57.

Smoot,G.F.(2007).Nobel Lecture:Cosmic microwave background radiation anisotropies:Their discovery and utilization.*Reviews of Modern Physics*,79(4),1349.

Stevenson,D.J.(2004).Exploring Mercury:The Iron Planet.

Tully,R.B.,Rizzi,L.,Shaya,E.J.,Courtois,H.M.,Makarov,D.I.,Jacobs,B.A.(2009).The extragalactic distance database. *The Astronomical Journal*,138(2),323.

Upgren,A.R.(2013).Night has a thousand eyes:a naked-eye guide to the sky,its science,and lore.*Springer*.

Zapolsky,H.S.(1968).Can the redshifts of quasi-stellar objects BE gravitational?*Astrophysical Journal*,vol.153, L163.

Zwicky,F.(1929).On the possibilities of a gravitational drag of light.*Physical Review*,34(12),1623

Revisiting Olbers' paradox with a modern lens

Abstract

Olbers' paradox has been used as an argument against the infinitude of the universe for more than 200 years. However, since its initial formulation, our perception of the universe has changed dramatically. In this article, Olbers' paradox is reviewed and compared with up-to-date observational data from the Principal Galaxies Catalogue (PGC). Trends in the number of galaxies and the total amount of light for increasing distance intervals may still suggest a finite universe based on Olbers' paradox. However, trends in the total angular area of galaxies with increasing distance intervals suggest that the universe as a whole is probably infinite. With this discrepancy in results among observational tests, the paradox remains mathematically unresolved. However, many of the original assumptions of the paradox should no longer be considered valid and an additional unstated assumption about the nature of light was challenged by the discovery of light quantization in the early 20th century. Ultimately, it is this unstated assumption of Olbers' paradox that suggests that the universe is infinite as at distances beyond about 330 Mpc, images of galaxies become invisible to the Hubble Space Telescope's (HST), let alone the naked eye. Therefore, there should no longer be any good reason to argue for the finitude of the universe based on Olbers' paradox.

Introduction

Early heliocentric scholars like Nicolaus Copernicus, Giordano Bruno and Thomas Digges believed in an infinite universe that had no center. One of the first astronomers to argue for a finite universe was Johannes Kepler in the early 1600s. Kepler suggested that the universe could not be infinite, for if it was, starlight would fill the entire background of the sky and make it at least as luminous as the Sun (Tessicini 2022). Similar arguments with more quantitative routes were presented in papers by Edmund Halley in the early 1700s (Halley 1720). Most astronomers in the early 18th century viewed the universe as an infinite sea of stars. In 1744, Loys de Cheseaux described a set of concentric shells of uniformly distributed stars around the Earth and concluded that at a distance of approximately 900 million Mpc, starlight would eventually fill the entire sky (Loys de Cheseaux 1744). To rectify what he calculated with observational reality, Loys de Cheseaux proposed the existence of an undiscovered light absorbing medium that permeates the universe and prevents a small fraction of starlight from reaching observers on Earth.

An essay written by Immanuel Kant in 1755 was influential in bringing modern ideas of solar system formation, the Milky Way galaxy, and the existence of other galaxies into the scientific noosphere (Kant 1755). This was followed in 1771 by the publication of Charles Messier's catalogue of nebulae, which further enhanced our understanding of galaxies (Messier 1771). By the late 18th century, many astronomers were interpreting the universe as a sea of island universes separated by expansive stretches of empty space rather than an infinite sea of stars. Then in 1823, Heinrich Wilhelm Olbers argued that the universe could not be infinite because if it were, we would see a galaxy along any possible line of sight we looked and therefore the background of the night sky would be completely filled with light (Olbers 1826). Nine years later, John Herschel argued against Loys de Cheseaux's light absorbing medium, for an infinite arrangement of stars would just heat that medium up to the point that it would disperse and prevent it from absorbing any light (Herschel 1831).

In 1901, Lord Kelvin proved that the Milky Way did not contain enough stars to cover the whole night sky. He then expanded on Olbers' paradox, stating that the universe must be finite not only in space but also in time as in an infinite universe, a telescope aimed in any direction of the night sky would end its line of sight on a distant galaxy from sometime in the universe's past (Thompson 1901). This was just one year after Max Planck published evidence for the quantization of light and four years before Albert Einstein published his work on the photoelectric effect (Planck 1900, Einstein 1905). It is Lord Kelvin's version of Olbers' paradox that endures as the preferred argument against an infinite universe (Harrison 1986, Newton 2001).

An alternative resolution to Olbers' paradox was presented by Carl Charlier in 1908, who proposed a hierarchical universe where increasing amounts of space contain decreasing densities of matter (Charlier 1908). Charlier's ideas did not receive much traction amongst the scientific community until the development of fractal cosmology many decades later (Pietronero 1987). After Edwin Hubble's telescopic observations and the discovery of Hubble's law in the 1920s, most astronomers adopted a belief of a finite, expanding universe. No one really bothered to challenge Olbers' paradox for many decades after this as giving the universe a finite size resolved the paradox.

Then came the QMAP, MAXIMA and BOOMER-ANG balloon experiments of the Cosmic Microwave Background Radiation (CMB) in the late 1990s, which reported essentially no large-scale spatial curvature in the universe, to an error bound of 0.4% (de Bernardis et al 2000, Biron 2015). These discoveries came as a surprise to most cosmologists as topologically speaking, flatness of a simply connected space implies infinitude (Ellis and Van Elst 1999, Tegmark 2015). As observational techniques improved and the large-scale distribution of galaxies in the universe became more well-defined, the universe's fractal dimension was determined to be approximately equal to 2, which has not helped in quantitatively resolving the paradox (Joyce et al 2005).

Mathematical Formulation

The essence of Olbers' paradox is expressed mathematically by the following equation.

$$l = \iiint_{r_0}^{\infty} E(r) \cdot B(r) \cdot N(r) \cdot r^2 sin\theta dr d\theta d\varphi$$
$$= 4\pi \int_{r_0}^{\infty} E(r) \cdot \frac{L(r)}{4\pi r^2} \cdot N(r) \cdot r^2 dr$$
$$= \int_{r_0}^{\infty} E(r) \cdot L(r) \cdot N(r) \cdot dr$$

l is the total amount of light the Earth receives from the whole sky per unit time.

r_0 is a distance that defines the first concentric shell.

$N(r)$ is a density function that shows the number of galaxies within an angular area of a spherical shell at distance r.

$B(r)$ is the average intensity of light on Earth for each galaxy at distance r.

$L(r)$ is the average luminosity of galaxies within a spherical shell, related to brightness by

$$B(r) = \frac{L(r)}{4\pi r^2}$$

$E(r)$ is a hypothetical light extinction function which could be equal to or slightly less than 1.

The mathematical aim of Olbers' paradox is to resolve whether the series represented by the integral is finite or infinite. If either $E(r)$, $L(r)$ or $N(r)$ decrease slightly with distance, then the integral eventually converges and the universe becomes intuitively infinite. In Loys de Cheseaux's formulation, he proposed the existence of a very sparse extragalactic medium that could block a small fraction of the light from distant galaxies. This essentially meant that Loys de Cheseaux made $E(r) < 1$ in order to force the integral to converge. Modern observations reveal, however, that the extragalactic medium performs very little absorption. Images of distant galaxies are seldom blurred and the extragalactic medium is extremely transparent.

Olbers' original 1823 paradox stated four assumptions:
1. The universe is homogenous, looking approximately the same in all directions.
2. The universe is unchanging in time when viewed on a large enough scale.
3. The universe is not undergoing any major systematic motions.
4. The laws of physics apply the same in all areas of the universe.

The first assumption, homogeneity, is not supported by modern astronomical observations. Both the large-scale distribution of galaxies and anisotropies in the CMB show that the universe is neither homogeneous nor isotropic on the largest observable scales. Instead, matter in the universe appears to have a clumpy structure and is denser in the general direction of the Shapley Supercluster (Hoffman et al 2017).

The Big Bang Theory suggests that Olbers' second assumption, that the universe is unchanging in time at large enough scales, is irrelevant. In a similar way, there is no reason to retain Olbers' third assumption of no large-scale motions occurring in the universe. The Milky Way and most galaxies in its vicinity are travelling towards the Great Attractor at ~250 km s[-1], which in turn is moving towards the Shapley Supercluster (Bolejko and Hellaby 2008). Our view of the universe from the vantage point of the Earth is certainly affected by these large-scale motions.

The fourth assumption of Olbers' paradox is still valid. Astronomical observations over the last two centuries have not provided any good reason to believe that the laws of physics would be different in other parts of the universe. Modern formulations of Olbers' paradox rely heavily on line-of-sight arguments, which say that in a hypothetically infinite universe, a telescope aimed in any direction will eventually meet the image of a galaxy from some time in the deep past.

Analysis

The analysis presented here uses the easily accessible and reliable Principal Galaxies Catalogue (PGC), which contains ~21300 principal galaxy groups with redshift-independent distance measurements (Tully et al 2009). The average luminosity of these principal galaxies is about 36×10^9 L⊙, which is comparable to the Andromeda galaxy (M31). Figure 4A below shows the distribution of these galaxies as a function of distance, arranged into 10 Mpc bins.

Figure 4A has two peaks, the first of which, between 20 and 30 Mpc, represents the latter part of the Messier catalogue's range. Beyond this is somewhat of a trough that lasts until ~110 Mpc and is followed by a pseudo-linear trend that leads up to a second peak at ~330 Mpc. This pseudo-linear trend can be approximated by the power function $\delta(r) = 0.4483r^{1.2757}$. Dividing this by $4\pi r^2$ gives a general density function $N(r) = 0.0357r^{-0.7243}$ that can be substituted into the total light integral. The strength of this pseudo-linear trend ($R^2 = 0.9744$) gives good reason to believe that not many principal galaxy groups are missing from the analysis in this particular distance range.

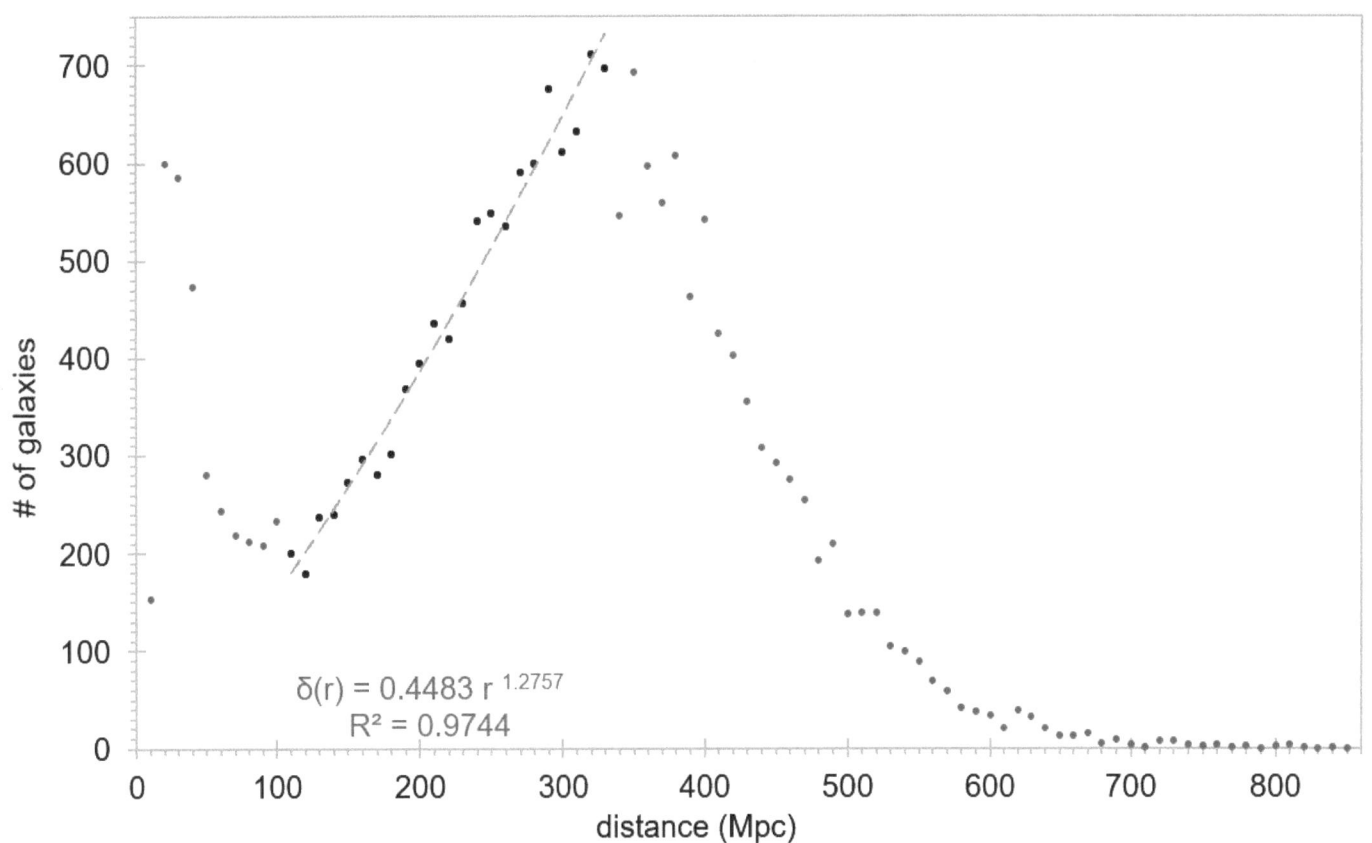

Figure 4A: The number of galaxies in the night sky, arranged into 10 Mpc concentric shells (Tully et al 2009). The equation of the dashed best fit function between 110 and 330 Mpc is shown.

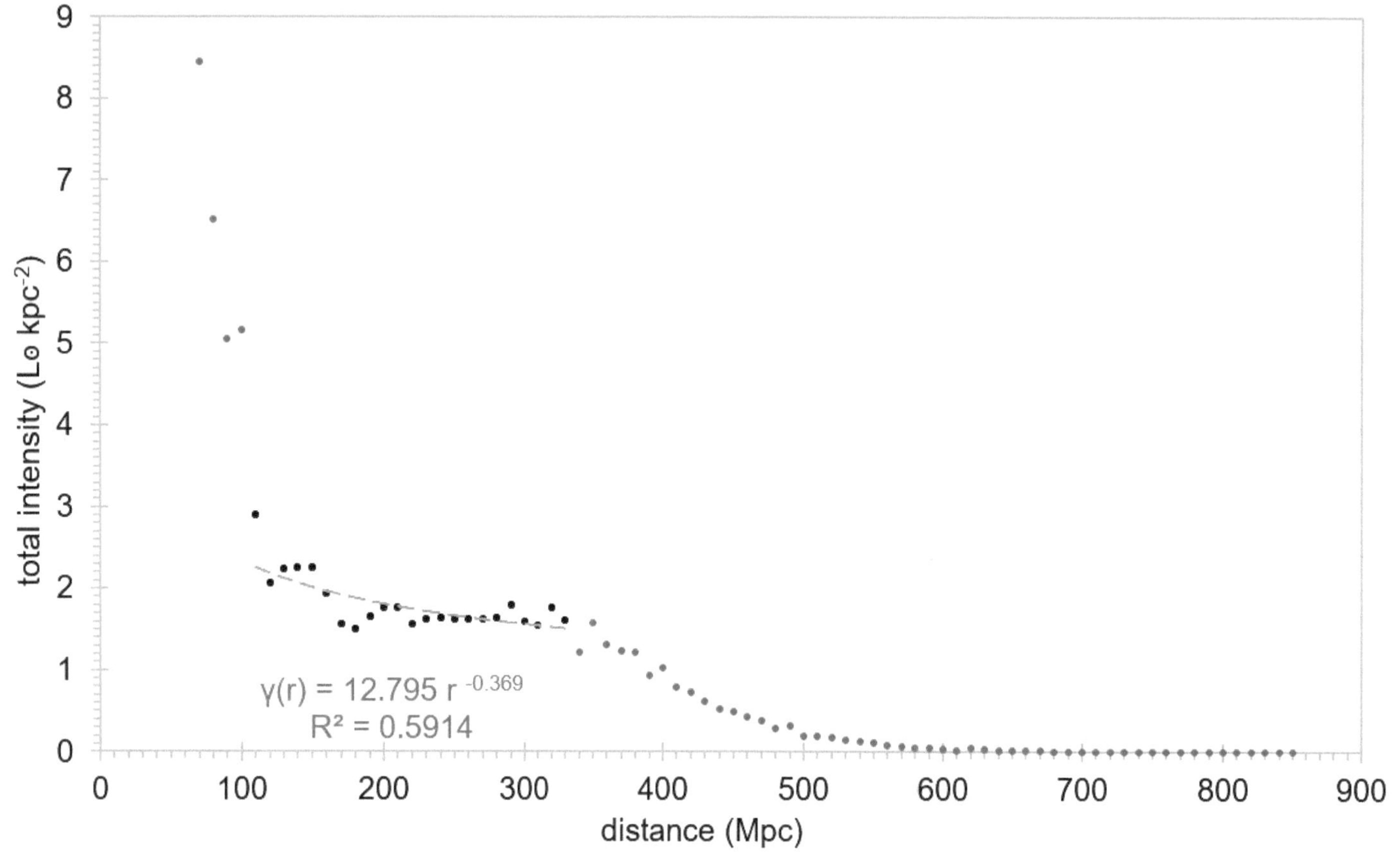

Figure 4B: The total amount of extragalactic light received by the Earth, arranged into 10 Mpc concentric shells beyond 60 Mpc (Tully et al 2009). The equation of the dashed best fit power function between 110 and 330 Mpc is shown.

The equation shown on the graph:

$$\gamma(r) = 12.795\ r^{-0.369}$$
$$R^2 = 0.5914$$

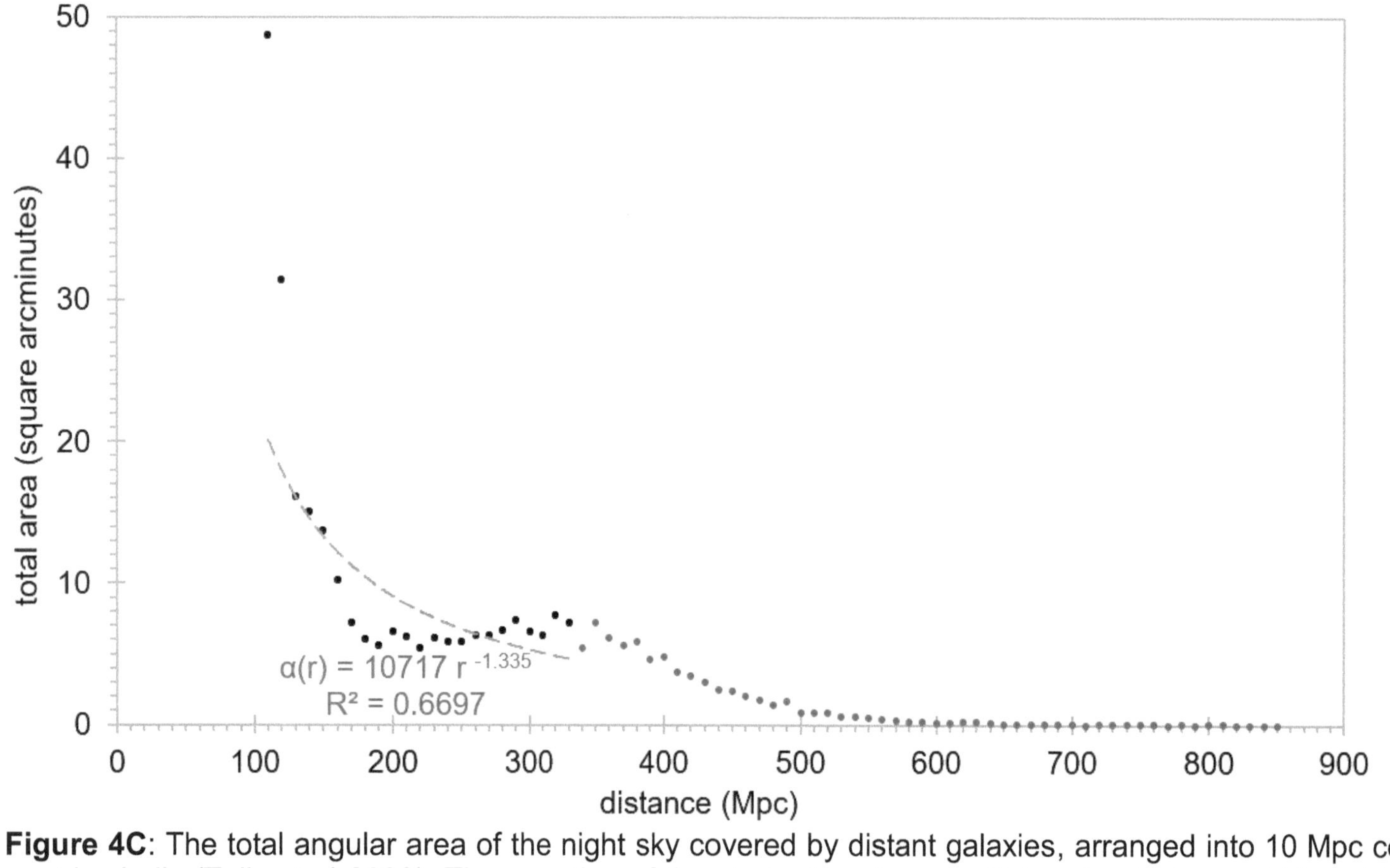

Figure 4C: The total angular area of the night sky covered by distant galaxies, arranged into 10 Mpc concentric shells (Tully et al 2009). The equation of the dashed best power fit function between 110 and 330 Mpc is shown.

The equation shown on the graph:

$$\alpha(r) = 10717\ r^{-1.335}$$
$$R^2 = 0.6697$$

26

The distribution of galaxies in figure 4A falls sharply after the second peak, which reflects the shortage of redshift-independent distance measurements beyond 330 Mpc. However, the strong pseudo-linear trend from 110 to 330 Mpc implies that the number of actual galaxies could keep increasing steadily as we move to further and further concentric shells.

If we substitute figure 4A's best fit density function and the average luminosity of principal galaxies into the integral, we get an infinite result.

$$l = \int_{r_0}^{\infty} E(r) \cdot L(r) \cdot N(r) \cdot dr$$

$$= \int_{110Mpc}^{\infty} 1 \cdot (3.6 \times 10^{10} L_{\odot}) \cdot 0.0357 r^{-0.7243} \cdot dr$$

$$= \infty$$

This result would prove that the universe must be finite. However, it would be irresponsible to conclude that as it would assume that distant galaxies beyond 330 Mpc will be visible to observers on Earth. In other words, it is warranted to assume that the number of galaxies in successive concentric shells beyond this distance will continue to increase, but it is not warranted to assume that beyond 330 Mpc astronomers will be able to find the same percentage of the galaxies that actually exist in these more distant concentric shells.

Figure 4B shows the total amount of light received on Earth by successive 10 Mpc concentric shells of galaxies. The collective light received by the 2334 principal galaxy groups in the Messier catalogue's range (d < 60 Mpc) is ~6.6×10^{-4} L⊙ pc⁻². This is roughly equivalent to the light received by the two Magellanic clouds and is too high to be shown on figure 4B's scale.

The remaining 19000 galaxies in the PGC send ~8.1×10^{-5} L⊙ pc⁻² to observers on Earth, which is hardly lighting up the night sky. The total light curve in figure 4B appears to flatten between 110 Mpc and 330 Mpc, the same range of distances as the pseudo-linear trend in figure 4A. Beyond 330 Mpc, the overall amount of light from galaxies with known non-redshift distances decreases sharply and consistently. With such a huge dropoff after 60 Mpc, it is hard to imagine how the addition of fainter and fainter galaxies will bring this total amount of light up substantially.

Despite this, integrating figure 4B's best fit power function beyond 110 Mpc yields an infinite amount of light.

$$l = \int_{110Mpc}^{\infty} 12.795 r^{-0.369} \cdot dr$$

$$= \infty$$

Figure 4C displays the total angular area of the night sky taken up by principal galaxies in successive 10 Mpc concentric shells. This chart is instrumental in gauging whether more and more concentric shells will contribute to eventually covering the entire night sky. The total angular area decreases sharply at first but then appears to flatten out after ~110 Mpc. A slight uptick between ~250 Mpc and ~320 Mpc seems to be somewhat of an anomaly rather than a general trend. As with the number of galaxies and the total light received, the angular area of principal galaxies with non-redshift derived distance measurements falls sharply beyond ~330 Mpc.

However, in contrast to figures 4A and 4B, integrating the best fit function of the total angular area does not yield an infinite result.

$$\alpha = \int_{110Mpc}^{\infty} 10717 r^{-1.335} \cdot dr$$

$$= 10717 \left[\frac{r^{-0.335}}{-0.335} \right]_{110}^{\infty}$$

$$= 6624 \text{ square arcminutes}$$

This only represents about 0.0045% of the night sky's total area. In contrast, the total angular area of the principal galaxies closer than 110 Mpc is ~26878 square arcminutes, or 0.018% of the night sky. Altogether, this is not even close to covering the whole celestial sphere. Therefore, in this analysis of Olbers' paradox, only 2 of the 3 mathematical tests are suggesting that the universe is finite.

With this information in mind, it seems obvious that the amount of extragalactic light the Earth receives from successive concentric shells sharply declines after 60 Mpc. This dramatic reduction suggests that even if the universe is infinite, the light from more and more distant galaxies doesn't accumulate enough to affect the night sky significantly.

27

Image extinction

An additional assumption has been required for the line-of-sight argument of Olbers' paradox to be valid. When formulated, a line of sight between an observer on Earth and a distant galaxy was assumed to represent an infinite stream of light energy, rather than a set of passing photons. The number of photons produced every second by a galaxy along a given line of sight is not infinite, although it is unimaginably numerous. Although this had been known by physicists since the early days of the 20th century, it has not brought the validity of Olbers' paradox into question very much in academic literature. It is somewhat curious why the paradox was still being used in the late 20th century to imply that the universe is finite (Wesson 1991).

Extragalactic image extinction does not mean that photons are terminated, but merely that the image of a galaxy can no longer be resolved by observers on Earth. Although photons from distant galaxies still reach the Earth, there may not be enough of them per second to generate a consistent image in the eye of an observer. This has the result of not causing the night sky to be lit up along a particular line of sight. Future telescopes may be able to peer further and further into the depths of extragalactic space and reveal more galaxies previously hidden to us but regions of the night sky will still remain empty and pitch black to naked eye observers on Earth. Don't forget that Olbers' original paradox stated that in an infinite universe, the night sky would appear bright like the day to a naked eye observer.

Since the amount of light travelling along a particular line of sight every second is finite and quantized, image extinction is related to effects of increasing distance on the human eye and today's most powerful telescopes. The human eye can detect light to a lower limit of ~5 to 7 photons per second, which places the limits of human vision at around 5.9×10^{-10} Wm^{-2} per steradian with a resolving power of approximately 28 arcseconds for white light at 550 nm (Hecht et al 1942, Sears et al 1982, Deering 1998). A relatively average galaxy like M31 has a luminosity of 3.8×10^{26} W (Van den Bergh 1999). With brightness following the inverse square law of luminosity, we can determine the upper limit of the distance that an average galaxy can be seen by naked eye observers using the equation below.

$$B(r) = \frac{L(r)}{4\pi r^2} = \left(\frac{5.9 \times 10^{-10}\ W}{m^2 radian^2}\right)\left(\frac{1\ radian}{206265\ arcseconds}\right)^2 \left(\frac{3.086 \times 10^{22}}{1\ Mpc}\right)^2$$

$$\frac{(3.8 \times 10^{26}\ W)}{4\pi r^2} = \frac{(1.3 \times 10^{25}\ W)}{Mpc^2 arcsecond^2}$$

$$r = 1.5\ Mpc$$

This calculation means that average galaxies like M31 will eventually become completely invisible to the naked eye beyond a distance of ~1.5 Mpc, even in the absence of the atmosphere. With its 2.4 m mirror and 57.6 m focal length, the Hubble Space Telescope (HST) has a resolving power of ~0.04 arcseconds (Chaisson and Villard 1990). This amounts to a magnification factor of around 700× in the optical wavelengths, extending the visible limit of average galaxies like M31 to ~1 Gpc. Not surprisingly, this is just beyond the limits of non-redshift derived distance measurements in the PGC.

At this point, it may be a worthwhile exercise to look at a Deep Field image taken by the HST, which is shown in figure 4D. The Deep Field offers a perspective that earlier thinkers such as Loys De Cheseaux, Olbers and Lord Kelvin did not have access to. In the Hubble Deep Field project, the telescope aimed at a particularly dark section of Ursa Major and collected light continuously for ten days (Ferguson 1998). The Deep Field allowed astronomers to peer further into the depths of extragalactic space than ever before, bringing into view an immense collection of galaxies of great multitude and diversity. Some of them may look very different now than when the light first left them billions of years ago. The Hubble Deep Field shows just how vast the universe is. Although most astronomers in the 1990s believed that the universe was finite, to many non-astronomers who are not well-versed in modern cosmological theories, the Hubble Deep Field conjures thoughts like "space just never ends" and "how can the universe be anything but infinite?"

Figure 4D: Hubble space telescope's deep field image. (obtained from *Hubblesite.org*)

The real point is that without large telescopes and extended-exposure photographic techniques, most galaxies in the Hubble Deep Field are completely invisible to observers on Earth. There's just not a thick enough stream of photons coming from those distant sources to make a consistent image. Figure 4D shows uncountable distant galaxies existing in what appears as a particularly dark piece of the night sky to naked eye observers. Such a conundrum reinforces the notion that we cannot think of these distant galaxies as suppliers of steady streams of light that allow us to see them no matter what distance they are at. Instead, at larger and larger distances the images of galaxies become fainter and fainter and eventually slip into an abyss of invisibility like Cantor dust. What would Olbers have to say about that if he were alive today?

Conclusions

The revisitation of Olbers' paradox presented here suggests that we should not consider the universe to be finite. Although Olbers' paradox passed two out of three observational tests, many of the original assumptions of Olbers' paradox no longer apply to our modern understanding of the universe. Olbers' paradox passed the first mathematical test, as the number of galaxies versus their distance increased with a divergent power function. This was also the case with the second test based on the total amount of light received. However, the third observational test related to the total angular area of the night sky illuminated by extragalactic light, was approximated by a best fit power function that caused the integral to converge rapidly. Therefore, the third observational test invalidated the line of sight arguments of Olbers' paradox. Since the origin of the paradox, our understanding of light and how vision works has evolved. An unstated assumption about unquantized light further hinders modern arguments for a finite universe according to Olbers' paradox.

This article has argued that images of distant galaxies eventually go extinct at large enough distances. The finitude of the photons emanating from distant galaxies eventually causes their images to become extinct beyond a set distance where they become invisible to observers on Earth. This argument is supported by an experimentally verified upper limit on naked eye vision of ~1.5 Mpc and an estimated upper limit on detection by the HST of ~1.0 Gpc. Lord Kelvin's assertion that any line of sight will eventually find a distant galaxy should no longer be considered valid in a universe where light is quantized.

This conclusion leads to a theoretically infinite universe, as more and more distant galaxies eventually become invisible to the human eye at large enough distances. The James Webb Space Telescope (JWST) will allow astronomers to peer even further into the cosmos and it will probably show that the universe just keeps going and going, and that our view of it just keeps extending as the resolving power of our telescopic instruments increases. Based on the analysis presented here, there should no longer be any reason at all to cling to the explanation of the finitude of the universe offered in Olbers' paradox.

References

Biron,L.(2015).Our universe is Flat.*Symmetry magazine.org. FermiLab/SLAC.*

Bolejko,K.,Hellaby,C.(2008).The great attractor and the shapley concentration.*General Relativity and Gravitation,40,* 1771-1790.

Chaisson,E.J.,Villard,R.(1990).The science mission of the Hubble Space Telescope.*Vistas in astronomy,33,*105-141.

Charlier,C.V.(1908).Wie eine unendliche Welt aufgebaut sein kann.*Meddelanden fran Lunds Astronomiska Observatorium Serie I,38,*1-15.

De Bernardis,P.,Ade,P.A.,Bock,J.J.,Bond,J.R.,Borrill,J.,Boscaleri,A.,Vittorio,N.(2000).A flat Universe from high-resolution maps of the cosmic microwave background radiation. *Nature,404*(6781),955-959.

Deering,M.F.(1998, May).The limits of human vision.In *2nd International Immersive Projection Technology Workshop* (Vol.2,1).

Einstein,A.(1905).On a Heuristic Viewpoint Concerning the Emission and Transformation of Light.*Annalen der Physik, 17.*

Ellis,G.F.,Van Elst,H.(1999).Cosmological models:Cargese lectures 1998.*Theoretical and Observational Cosmology,* 1-116.

Ferguson,A.S.(1998).The Hubble deep field (Les Champs profonds de Hubble).*Reviews in Modern Astronomy,11,* 83-115.

Halley,E.(1720).V.Of the infinity of the sphere of fix'd stars. *Philosophical Transactions of the Royal Society of London,31*(364),22-24.

Harrison,E.(1986).Kelvin on an old, celebrated hypothesis. *Nature,322*(6078),417-418.

Hecht,S.,Shlaer,S.,Pirenne,M.H.(1942).Energy,quanta,and vision.*The Journal of general physiology,25*(6),819-840.

Herschel,J.F.W.(1831).A preliminary discourse on the study of natural philosophy(Vol.1).*Longman,Rees,Orme,Brown, Green and Taylor.*

Hoffman,Y.,Pomarède,D.,Tully,R.B.,Courtois,H.M.(2017).The dipole repeller.*Nature Astronomy,1*(2),0036.

Joyce,M.,Labini,F.S.,Gabrielli,A.,Montuori,M.,Pietronero,L. (2005).Basic properties of galaxy clustering in the light of recent results from the Sloan Digital Sky Survey.*Astronomy & Astrophysics,443*(1),11-16.

Kant,I.(1755).Universal Natural History and Theory of the Heavens:Or, An Essay on the Constitution and the Mechanical Origins of the Entire Structure of the Universe.

Loys de Chéseaux,J.P.,Cassini de Thury,C.F.,Calandrini,J.L. (1744).Traité de la comète:qui a paru en décembre 1743 & en janvier,février & mars 1744.

Messier,C.(1771).Catalogue des nébuleuses et des amas d'étoiles,que l'on decouvre parmi les etoiles fixes.*Mem. Mat.Phys.Acad.des Sci.for.*

Newton,D.(2001).Olbers' Paradox:A Review of Resolutions to this Paradox.

Olbers,W.(1826).Über die Durchsichtigkeit des Weltraums (About the transparency of the universe).*Bode Jb,15.*

Pietronero,L.(1987).The fractal structure of the universe: Correlations of galaxies and clusters and the average mass density.*Physica A:Statistical Mechanics and its Applications,144*(2-3),257-284.

Planck,M.(1900).On an Improvement of Wien's Equation for the Spectrum.*Ann. Physik,1,*719-721.

Sears,F.W.,Zemansky,M.W.,Young,H.D.(1982).*University physics* (Vol.6).Reading, MA:Addison-Wesley.

Tegmark,M.(2015).Our mathematical universe:My quest for the ultimate nature of reality.*Vintage.*

Tessicini,D.(2022).Kepler's De Stella Nova(1606)on the nature and motions of the celestial novelties.*Kepler's De Stella Nova (1606) on the nature and motions of the celestial novelties,*292-292.

Thompson,W.(1901).XII.On ether and gravitational matter through infinite space.*The London, Edinburgh, and Dublin Philosophical Magazine and Journal of Science,2*(8),161-177.

Tully,R.B.,Rizzi,L.,Shaya,E.J.,Courtois,H.M.,Makarov,D.I.,Jacobs,B.A.(2009).The extragalactic distance database.*The Astronomical Journal,138*(2),323.

Van den Bergh,S.(1999).The local group of galaxies.*The Astronomy and Astrophysics Review,9*(3),273-318.

Wesson,P.S.(1991).Olbers's paradox and the spectral intensity of the extragalactic background light.*The Astrophysical Journal,367,*399-406.

Did life in our solar system start from scratch?

Abstract

The Miller-Urey experiment in the 1950s simulated atmospheric and surface conditions of the early Earth and proved that amino acids and lipids, the building blocks of life, could be created from simple compounds such as H_2O, NH_3 and CH_4 in the presence of an electric current. The results of this experiment contributed to a general view that life on Earth originated and evolved through purely endogenic processes, and exists as a sort of independent bubble in an ocean of lifelessness. This view was challenged by a meteorite that fell near the town of Murchison, Australia in 1969, which was found to contain an abundance of complex organic molecules including amino acids, purines and pyrimidines. This event opened up the possibility that amino acids, nucleic acids and other complex organic compounds could have been transported to the early Earth from outside. If this was the case, then basic Miller-Urey processes wouldn't be required for the genesis of life on Earth because the building blocks of life could have already been present on its surface. In this article, three different interpretations of the Murchison meteorite's origin are explored: a rogue comet, a previous solar system and abiotic amino acid production. Based on the analysis presented in this article, it is likely that the organic molecules on the Murchison meteorite represent the remnants of life from an ancient planet that existed before our solar system formed.

Introduction

The notion that life exists on other planets has been an enduring source of speculation and wonder since antiquity. The theory that life on Earth could have originated via external agents is called panspermia. In 1834, Jons Jacob Berzelius proposed the possibility that comets could carry organic molecules and act as vehicles that transport the seeds of life to young planets (Berzelius 1834). Similar ideas were presented by Lord Kelvin, Svante Arrhenius, Fred Hoyle and Chandra Wickramasinghe (Thompson 1871, Arrhenius 1908, Hoyle and Wickramasinghe 1981). An experiment performed by Stanley Miller and Harold Urey in 1952 provided results that contrasted greatly with earlier panspermic notions. When they simulated the surface conditions of the early Earth in a glass beaker, complex biomolecules formed, demonstrating that life could have originated abiogenetically without any deliveries from outside (Miller 1953). In 1969, a meteorite fell from the sky near the town of Murchison, Australia. Laboratory analysis of pieces of this meteorite, such as the one shown in figure 5A, revealed the presence of amino acids and other complex organic molecules (Kvenvolden et al 1970).

Figure 5A: A piece of the Murchison meteorite held at the Freiburg Naturmuseum.

In 1944, Fred Hoyle published the now widely accepted theory that our solar system formed from the aftermath of a supernova (Hoyle 1944). This theory became popular amongst astronomers over the following decades and was supported experimentally when studies of very old meteorites showed an abundance of ^{60}Fe and ^{26}Al, two isotopes that are produced almost exclusively in supernovae (Williams and Cremin 1968, Cameron and Truran 1977, Dauphas et al 2008). In the

1970s, Chandra Wickramasinghe performed observations on interstellar dust and found that it contained significant amounts of carbon and organic molecules (Wickramasinghe and Allen 1980). Some more recent studies have even claimed that molecules such as glycine, an amino acid, could exist in interstellar space (Kuan et al 2003).

Data

In total, about 100 kg of the Murchison meteorite's material was recovered and analyzed. Laboratory results showed that it contained a high concentration of organic molecules, including several types of amino acids, purines and pyrimidines, as shown in figure 5B (Botta and Bada 2003). Isotopic analysis and the fact that it was an observed fall proved that these organic molecules were of extraterrestrial origin and not the result of contamination after the meteorite landed. Radiometric dating of silicon carbide particles within the Murchison meteorite revealed an age of ~7 Gyr, which is higher than the estimated age of the solar system: 4.57 Gyr (Bouvier and Wadhwa 2010, Heck et al 2020).

Purines and pyrimidines detected on the Murchison meteorite included all five of the nucleic acids found in DNA and RNA on Earth (Oba et al 2022). The Murchison meteorite's purines and pyrimidines are definitively not terrestrial in origin as they have $\delta^{13}C$ ratios between +44.5‰ and +37.7‰, as opposed to -10‰ to -33‰ for the vast majority of organic carbon on Earth (O'Leary 1988).

A further surprise came when most of the meteorite's amino acids were found to be racemic (Kvenvolden 1970). This means that they were of mixed chirality, with approximately 50% of the amino acids being left-handed and the other 50% of them being right-handed (Schmitt-Kopplin et al 2010). Left-handed and right-handed isomers appear as mirror images to each other, as illustrated in figure 5C. Amino acid isomers have identical physical and chemical properties, but almost all protein chains on Earth are composed of left-handed amino acids. This contrasts with amino acids produced abiotically in laboratory settings, which are usually racemic (Bada and McDonald 1996, Zubay 2000).

substance	concentration (ppm)	notes
iron	221000	
water	120000	
aromatic hydrocarbons	3319	
carboxylic acids	>300	
fullerenes	>100	
sulphonic acids	68	
amino acids	17 – 60	70 compounds
aliphatic hydrocarbons	35	
hydrocarboxylic acids	15	
alcohols	11	
phosphonic acids	2	
purines & pyrimidines	1.3	$38‰ < \delta^{13}C < 45‰$

Figure 5B: Chemical constituents of the Murchison meteorite (Machalek 2007).

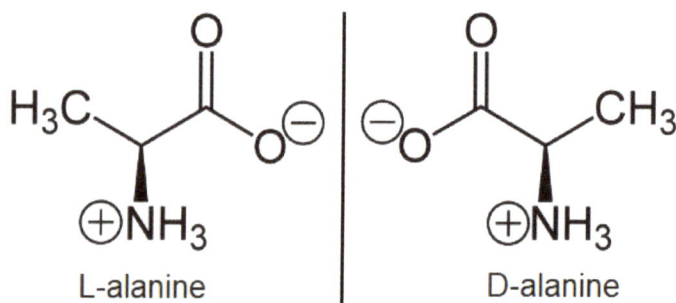

L-alanine   D-alanine

Figure 5C: Isomers of alanine.

When biological enzymes catalyze reactions, they enforce homochirality on organic molecules (Seckbach 2012). Right-handed amino acids are rare on Earth but it's not impossible that some extraterrestrial ecosystems could contain mostly right-handed amino acids. In a similar way, the vast majority of sugars, including glucose and deoxyribose, in living cells on Earth are right-handed isomers (Zubay 2000). In the absence of biological processes, homochiral assemblages of amino acids are gradually converted into racemic ones. Whenever a free radical substitution reaction occurs at a chiral carbon, racemization is almost always observed (Smith and March 2001). Racemization periods vary among different amino acids and depend on temperature. They range from ~1 to ~2 Myr at a temperature of -10°C and would be longer in the cold space beyond Earth's orbit (Miller et al 2013). With this in mind, it's possible that the Murchison meteorite's amino acids could represent remnant biological proteins that existed long ago, but eventually racemized after being exposed to UV radiation and cosmic rays for several billion years.

Three possible scenarios regarding the nature of the Murchison meteorite are worth discussing.

1. That the Murchison meteorite came from a far away ancient planet that had life.
2. That the Murchison meteorite came from a life-bearing planet that existed in our vicinity before the supernova that initiated the solar system's formation.
3. That the organic molecules on the Murchison meteorite formed abiotically in outer space.

We will explore each of these possibilities in turn.

Rogue interstellar comet

It's possible that the Murchison meteorite could represent a piece of a rogue asteroid or comet that wandered through interstellar space for billions of years until stumbling into the Earth's gravitational pull. In the last decade, a number of interstellar comets and meteorites have been discovered including 1I/Oumuamua and 2I/Borisov (Meech et al 2017). Therefore, it is quite possible that the Murchison meteorite could have come from outside the solar system. It's even possible that a nearby star may have gone supernova, sending pieces of its planets flying in all directions. As the Murchison meteorite contains amino acids, whatever planet it may have come from could have been life-bearing and abiotic processes could have gradually converted the originally homochiral mixture of amino acids into a racemic mixture over the meteorite's long journey through cold interstellar space.

The previous solar system

Another possible explanation for the Murchison meteorite is that it represents a piece of rock from the planetary system of our Sun's parent star. Since our solar system probably formed from the remnant of a supernova, at least one star must have existed in our pre-solar neighbourhood. When that parent star's lifetime ended in a supernova, small pieces of any planets it had could have ended up in our solar system's protoplanetary cloud. Given that so far more than 3900 exoplanetary systems have been detected and more than 850 of them contain more than one planet, it is not unlikely that our Sun's parent star had planets, and that one or more

of those planets could have had life (Schneider 2010).

Meteorites containing complex biomolecules, remnants of whatever life forms may have inhabited this previous planetary system, may have even ended up on the surface of the primordial Earth. Meanwhile, the particular piece of rock that would eventually become the Murchison meteorite may have initially avoided capture by any of our solar system's protoplanets and instead wandered undisturbed until eventually being captured by Earth's gravitational field billions of years later. Think about it; if at least one meteorite with complex organic molecules has landed on Earth in the last century, imagine how many similar pieces of rock could have fallen on the Earth's surface in its first few million years after formation.

Although the creation of large biomolecules could have happened abiogenetically on the early Earth through Miller-Urey processes, life on Earth may not have needed to start with such minimal initial conditions as complex organic molecules may have already been present on the Earth's surface. The Murchison meteorite may in fact show us that the transportation of biochemical remnants from a previous planetary system to our planet is an ongoing process.

Abiotic interstellar molecules

The possible detection of glycine in interstellar gas clouds suggests that the building blocks of life may exist abiotically in outer space (De Jesus et al 2021). In fact, it is possible for amino acids to form from random collisions between interstellar atoms. Carbon-carbon bonds are very durable in low-energy conditions that lack a strong source of radiation. This can lead to large collections of carbon atoms, such as C_{60} and C_{70}, simply floating around in interstellar space with no biological purpose (Cami et al 2010). Is it possible that amino acids and other complex organic molecules formed abiotically on the Murchison meteorite while it was travelling through space?

Some moon rocks collected by the Apollo astronauts contained trace amounts of amino acids, up to 70 parts per billion (Fox 1973). However, it was later found through isotopic analysis that these lunar amino acids had $\delta^{13}C$ values between -33‰ and -22‰, which suggested that these amino acids entered the rocks via terrestrial contamination

(Brinton and Bada 1996, Elsila et al 2016). Despite this, the lunar rocks also contained trace amounts of two non-proteinogenic amino acids, 2-aminoiso-butyric acid and α-aminobutyric acid, which suggested that a small portion of the lunar amino acids could have formed abiotically due to HCN implantation on lunar soil under the action of solar wind (Brinton and Bada 1996, Elsila et al 2016). If this is true, that amino acids can indeed form on the surface of the moon, it is not a large stretch to suggest that they could form on the surface of meteorites. However, the large difference between the amino acid concentration of the Murchison meteorite (17-60 ppm) and the non-proteinogenic amino acid concentration of lunar rocks (<70 ppb) casts serious doubt on this interpretation.

Discussion

The third scenario, abiotic origin, is unfavourable as although abiotic generation of amino acids is possible on meteorites, the Murchison sample has a much higher concentration of amino acids than any known extraterrestrial abiotic sample. For that reason, it is more likely that the Murchison meteorite's amino acids represent remnants of ancient biological processes from a time before the solar system's formation. Given that the Murchison sample is such an anomaly among meteorites, it being the only meteorite with an observed fall that contains such high concentrations of organic molecules, our ability to know whether it represents a one-time-only event, or one with recurring frequency, is hindered. With this in mind, the following interpretation is offered. If the Murchison meteorite is truly anomalous then it probably represents a rogue interstellar comet. If more Murchison-type meteorites are added to the record, then it would suggest that they are remnants of the planetary system of our Sun's parent star.

Conclusions

Due to a number of advances in astronomy in recent decades, a better understanding of what the 1969 Murchison meteorite represents has become possible. Although it is an anomaly among meteorites, it suggests that the building blocks of proteins and DNA could have been transported to the early Earth from elsewhere in the cosmos. It also shows that remnant biomolecules from extraterrestrial life could be continuously pelting the Earth over centuries and millennia and that the early Earth may not have had to manufacture the building blocks of life from scratch.

After considering the facts and scenarios presented here, it seems likely that organic chemicals within the Murchison meteorite are remnants of biologically derived molecules that came from a life-bearing planet blown apart by a supernova. It's possible that a larger number of Murchison-type meteorites could have been present in the vicinity of the early Earth when our solar system was forming from its primordial nebula, a previous star's supernova remnant. Although abiogenesis is possible, it wouldn't have been necessary if amino acids didn't have to be built from scratch on the early Earth.

Much like an isolated, newly formed volcanic island in a vast ocean, in its infancy the Earth had all the necessary conditions for life to exist but needed some time for a spark of life to come along. Complex organic molecules like amino acids and nucleic acids may in fact be hitching rides on comets on lengthy cosmic journeys to distant planets just waiting to be fertilized with the seeds of life. It makes one wonder whether life on Earth could be part of a much greater legacy than just the ~4 Gyr evolutionary story told in the geological record. Perhaps our planet's biochemistry is part of a much larger family tree with roots extending deep into the cosmos.

References

Arrhenius,S.(1908).Worlds in the making: the evolution of the universe.*Harper&brothers*.

Bada,J.L.,McDonald,G.D.(1996).Peer reviewed:detecting amino acids on Mars.*Analytical chemistry*,68(21),668A-673A.

Berzelius,J.J.(1834).Analysis of the Alais meteorite and implications about life in other worlds.*Ann Chem Pharm*,10(1),134-135.

Botta,O.,Bada,J.L.(2002).Extraterrestrial organic compounds in meteorites.*Surveys in Geophysics*,23,411-467.

Bouvier,A.,Wadhwa,M.(2010).The age of the Solar System redefined by the oldest Pb-Pb age of a meteoritic inclusion. *Nature geoscience*,3(9),637-641.

Brinton,K.L.,Bada,J.L.(1996).A reexamination of amino acids in lunar soils:Implications for the survival of exogenous organic material during impact delivery.*Geochimica et cosmochimica acta*,60(2),349-354.

Cameron,A.,Truran,J.(1977).The supernova trigger for formation of the solar system.*Icarus*,30(3),447-461.

Cami,J.,Bernard-Salas,J.,Peeters,E.,Malek,S.E.(2010).Detection of C_{60} and C_{70} in a young planetary nebula.*Science*,329(5996),1180-1182.

Dauphas,N.,Cook,D.L.,Sacarabany,A.,Froehlich,C.,Davis,A.M.,Wadhwa,M.,Gallino,R.(2008).Iron 60 evidence for early injection and efficient mixing of stellar debris in the protosolar nebula.*The Astrophysical Journal*,686(1),560.

De Jesus,D.N.,Da Silva,J.M.,Tejero,T.N.,De Souza Machado,G.,Xavier Jr,N.F.,Bauerfeldt,G.F.(2021).Chemical mechanism for the decomposition of CH_3NH_2 and implications to interstellar glycine.*Monthly Notices of the Royal Astronomical Society*,501(1),1202-1214.

Elsila,J.E.,Callahan,M.P.,Dworkin,J.P.,Glavin,D.P.,McLain,H.L.,Noble,S.K.,Gibson Jr,E.K.(2016).The origin of amino acids in lunar regolith samples.*Geochimica et Cosmochimica Acta*,172,357-369.

Fox,S.W.(1973).The Apollo Program and Amino Acids.*Bulletin of the Atomic Scientists*,29(10),46-51.

Heck,P.R.,Greer,J.,Kööp,L.,Trappitsch,R.,Gyngard,F.,Busemann,H.,Wieler,R.(2020).Lifetimes of interstellar dust from cosmic ray exposure ages of presolar silicon carbide.*Proceedings of the National Academy of Sciences*,117(4),1884-1889.

Hoyle,F.(1944).On the origin of the solar system.In *Mathematical Proceedings of the Cambridge Philosophical Society*(Vol.40,No.3,256-258).Cambridge University Press.

Hoyle,F.,Wickramasinghe,C.(1981).Comets-a vehicle for panspermia.In *Comets and the Origin of Life* (227-239).Springer,Dordrecht.

Kuan,Y.J.,Charnley,S.B.,Huang,H.C.,Tseng,W.L.,Kisiel,Z.(2003).Interstellar glycine.*The Astrophysical Journal*,593(2),848.

Kvenvolden,K.,Lawless,J.,Pering,K.,Peterson,E.,Flores,J.,Ponnamperuma,C.,Moore,C.(1970).Evidence for extraterrestrial amino-acids and hydrocarbons in the Murchison meteorite.*Nature*,228(5275),923-926.

Machalek,P.(2007).Organic Molecules in Comets and Meteorites and Life on Earth.*Department of Physics and Astronomy*.Johns Hopkins University.

Meech,K.J.,Weryk,R.,Micheli,M.,Kleyna,J.T.,Hainaut,O.R.,Jedicke,R.,Chastel,S.(2017).A brief visit from a red and extremely elongated interstellar asteroid.*Nature*,552(7685),378-381.

Miller,G.H.,Kaufman,D.S.,Clarke,S.J.(2013).Amino Acid Dating.In *Encyclopedia of Quaternary Science:Second Edition*(37-48).Elsevier Inc.

Miller,S.L.(1953).A production of amino acids under possible primitive earth conditions.*Science*,117(3046),528-529.

Oba,Y.,Takano,Y.,Furukawa,Y.,Koga,T.,Glavin,D.P.,Dworkin,J.P.,Naraoka,H.(2022).Identifying the wide diversity of extraterrestrial purine and pyrimidine nucleobases in carbonaceous meteorites.*Nature communications*,13(1),2008.

O'Leary,M.H.(1988).Carbon isotopes in photosynthesis.*Bioscience*,38(5),328-336.

Schmitt-Kopplin,P.,Gabelica,Z.,Gougeon,R.D.,Fekete,A.,Kanawati,B.,Harir,M.,Hertkorn,N.(2010).High molecular diversity of extraterrestrial organic matter in Murchison meteorite revealed 40 years after its fall.*Proceedings of the National Academy of Sciences*,107(7),2763-2768.

Schneider,J.(2010).Interactive extra-solar planets catalog.*The Extrasolar Planets Encyclopaedia*.

Seckbach,J.,ed.(2012).Genesis-in the beginning:precursors of life,chemical models and early biological evolution.*Dordrecht:Springer*.

Smith,M.,March,J.(2001).*March's advanced organic chemistry:reactions mechanisms and structure*(5th ed.).Wiley.

Thompson,W.(1871).ART.XXXVI.-Inaugural Address before the British Association at Edinburgh,August 2d.*American Journal of Science and Arts(1820-1879)*,2(10),269.

Wickramasinghe,D.T.,Allen,D.A.(1980).The 3.4-μm interstellar absorption feature.*Nature*,287(5782),518-519.

Williams,I.P.,Cremin,A.W.(1968).A survey of theories relating to the origin of the solar system.*Quarterly Journal of the Royal Astronomical Society*,9,40.

Zubay,G.(2000).Origins of life:on earth and in the cosmos.*Elsevier*.

Asteroid belts are common in the universe

Abstract

The existence of an asteroid belt between the orbits of Mars and Jupiter is not an accident. It is predetermined by the fact that abundant light molecules such as H_2, He, H_2O, NH_3 and CH_4 can accumulate in the atmospheres of planets that lie beyond our solar system's frost line. This allows gas giant planets to develop beyond a certain distance from the Sun, and these planets in turn exert strong tidal influences on smaller rocky bodies in the sunward direction. This consequently prevents major planetary bodies from forming in a region proximally inward of the frost line, which is why our solar system has a collection of asteroids between ~2 and ~4 AU instead of a single orbiting planet. The underlying astrophysical principles related to our solar system's frost line and debris belt formation likely apply to most other planetary systems, which would mean that asteroid belts are a relatively common feature in the universe.

Introduction

After the development of Kepler's laws of planetary motion and Newton's law of universal gravitation, accurate estimates of planetary distances from the Sun could be deduced. In the early 18th century, a general mathematical relationship emerged that approximated the location of the first six planets within the solar system. Now known as the Titius-Bode Law, its most common form was written as $a = 0.4 + 0.3 \times 2^n$ (Gregory 1715). In this formula, a is the distance of a planet from the Sun in astronomical units and n is the integer counter sequence $n = -\infty, 0, 1, 2 ...$

Figure 6A shows an overview of the Titius-Bode relationship and its relation to the planets in the solar system. The first four values in the law correspond with the orbits of the solar system's four inner planets, while Jupiter and Saturn correspond with $n = 4$ and $n = 5$. Although the absence of a planet at the $n = 3$ position left a gap in the sequence, the discovery of Uranus by William Herschel in 1781, at an orbital distance close to the formula's prediction for $n = 6$, instilled further confidence in the Titius-Bode Law (Herschel 1781).

Then on New Year's Day of 1801, Italian astronomer Giuseppe Piazzi discovered the dwarf planet Ceres orbiting at a distance of ~2.8 AU, which was within 1.2% of the value predicted by the Titius-Bode law for the $n = 3$ position (Piazzi 1801). Three other dwarf planets: Pallas, Juno and Vesta, were discovered by 1807, with orbits similar to Ceres (von Zach 1806). We now know of millions of small planetesimals that exist in a sparse ring between ~2 and ~4 AU known as the asteroid belt, illustrated in figure 6B on the next page (Junfeng et al 2016).

planet	semimajor axis (AU)	Titius-Bode # (n)	distance (AU)	error (%)	Richardson # (m)	distance (AU)	error (%)
Mercury	0.387	$-\infty$	0.4	+3.32	-2	0.387	0.00
Venus	0.723	0	0.7	-3.33	-1	0.723	0.00
Earth	1.000	1	1.0	0.00	0	1.000	0.00
Mars	1.524	2	1.6	+4.77	1	1.524	0.00
Ceres	2.769	3	2.8	+1.16	2	2.67	-0.04
Jupiter	5.204	4	5.2	-0.05	3	5.20	0.00
Saturn	9.583	5	10.0	+4.42	4	9.55	0.00
Uranus	19.22	6	19.6	+1.95	5	19.23	0.00
Neptune	30.07	7	38.8	+29.03	6	30.13	0.00
Pluto	39.48				7	41.8	+0.06

Figure 6A: Planetary orbital distances compared to values predicted by the Titius-Bode Law and the Richardson formulation.

When Neptune was discovered in 1846, its orbital distance was almost 30% larger than the value predicted for the $n = 7$ position (Le Verrier 1846). This caused the Titius-Bode Law to lose its designation as a general rule for the solar system. More refined variations of Titius-Bode-like relationships were produced by Mary Blagg in 1913 and D. E. Richardson in 1945, which more accurately model the semi-major axes of all eight planets in the solar system (Blagg 1913, Richardson 1945). The Richardson relation, also outlined in figure 6A, is described by the equation $R_n = (1.728)^m \varrho_m(\Theta_m)$, where m represents the integer counter sequence $m = -2, -1, 0, 1, 2, \ldots$ and $\varrho_m(\Theta_m)$ is an oscillatory function representing distances from an off-centered origin to angularly varying points on an ellipse. Blagg-Richardson formulations have also been used to accurately describe the semi-major axes of the satellites of Jupiter, Saturn and Uranus (Blagg 1913, Harwit 1988). A 2013 study of 68 exoplanetary systems with four or more planets showed that 96% of them abided by a Titius-Bode-type relationship (Poveda and Lara 2008). The number of planets so far detected in these exoplanetary systems is lower than expected but there may still be many Mars, Mercury or Ceres-sized planets in these systems that are too small for our current instruments to detect (Huang and Bakos 2014).

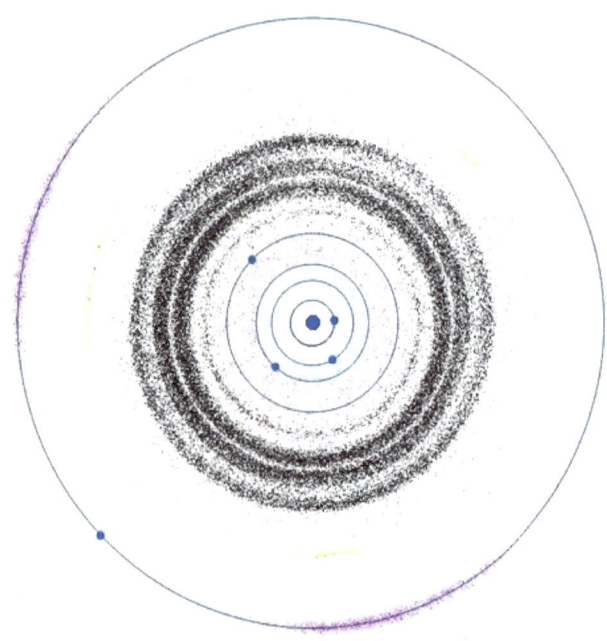

Figure 6B: Our solar system's asteroids. Jupiter's orbit is the outermost circle. Black dots show the main asteroid belt. Yellow dots show the Hilda asteroids and violet dots represent Jupiter trojans.

The Richardson formulation is able to describe the layout of our solar system in a rather ordered way rather than a random arrangement. Therefore, the location of the planets in our solar system is likely guided by some as-yet-not-fully-understood underlying astrophysical principles. If this is true, then the idea that set astrophysical rules and mechanisms play a lead role in determining the structure of the solar system can naturally be extended to exoplanetary systems in general. This in turn leads one to suppose that many exoplanetary systems might have an asteroid belt similar to our own.

The Frost Line

The frost line is a set distance from a star beyond which light molecules like H_2, He, H_2O, CH_4 and NH_3 can accumulate easily in planetary atmospheres. These molecules, which are among the most abundant in the universe, have different condensation temperatures, and therefore different frost lines, but most mentions of the frost line in scientific literature refer to the distance at which H_2O can accumulate in large quantities in planetary atmospheres. The H_2O frost line during the solar system's formation was probably somewhere between ~2.7 and ~3.2 AU, which overlaps the region that divides the four rocky inner planets and the four giant outer planets (Hayashi 1981, Kaufmann 1987, Podolak and Zucker 2004, Jewitt et al 2007, Martin and Livio 2012). At ~5 AU, Jupiter's orbital radius is almost double this value and would have had no trouble accumulating an atmosphere of H_2O, CH_4 and NH_3 in the early solar system. The solar system's current H_2O frost line may actually be closer to ~5 AU (Jewitt et al 2007).

Figure 6C provides a visual representation of the escape velocities and temperatures of various planetary bodies. The two main factors for determining a planet's atmospheric composition are mass and temperature. Mass is important as a stronger gravitational field increases the escape speed. Larger planets pull harder on gas molecules and prevent them from escaping. Temperature is important because hotter gas molecules have more kinetic energy and can move faster to escape the planet's gravity. If a gas molecule's speed exceeds the escape velocity, then it has a chance of leaving the atmosphere. Figure 6C is based on $v_{esc} < 10\ v_{rms}$, which means that roughly one in ten molecules that exceed escape velocity will leave the atmosphere.

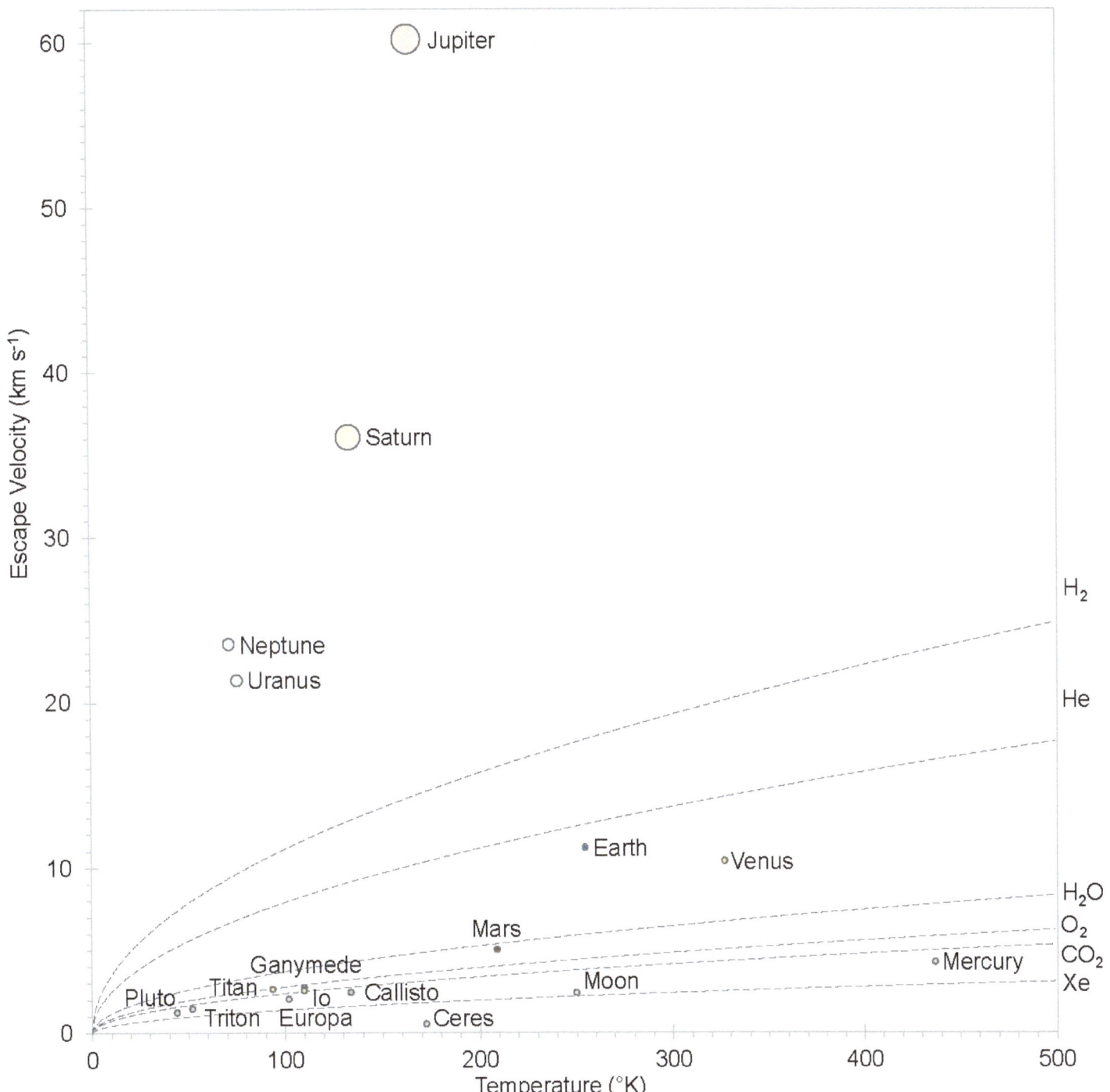

Figure 6C: Escape velocity curves for various gases, compared to bodies in the solar system.

In our solar system, the frost line lies somewhere beyond Mars's orbit, which has allowed Jupiter, Saturn, Uranus and Neptune to hold large amounts of H_2, He and H_2O in their atmospheres since the formation of the solar system. As you can see in figure 6C, the four gas giants can even retain H_2 and He in their atmospheres but their moons, such as Ganymede and Titan can not.

The location of our solar system's asteroid belt between ~2 and ~5 AU, immediately inward of the frost line, is a direct result of Jupiter's strong tidal forces repeatedly acting on this zone. Only dwarf planets can exist in this region, the largest of which has a diameter less than a third of the Moon's. Even if a large protoplanet once resided in this region, Jupiter's tidal forces would have just ripped it apart, leaving a field of rocky debris in its wake. It's in fact most likely that a large planet never existed there as Jupiter's tidal forces would have been acting on the region since the earliest days of the solar system.

Orbital stability

Titius-Bode type relationships, or Blagg-Richardson relationships, are not completely understood theoretically but likely have something to do with certain harmonic values that existed in the early stages of solar system formation, when protoplanets were accreting material and forming a set of orbits that were mutually stable. Our solar system has eight traditional planets as well as two debris belts, the asteroid belt and the Kuiper belt, that are all near integer values in the Richardson formulation. The Blagg-Richardson value of $n = 2$ corresponds with 2.67 AU. This does not match the orbit of Ceres, the largest asteroid. Instead, it lies near the orbit of Juno, a well-known smaller asteroid that resides in a very central part of the belt. The distribution of matter in the asteroid belt is partially determined by the Kirkwood gaps, empty bands within the belt that have orbital resonances with Jupiter's orbit. As you can see in figure 6D, particularly prominent gaps exist at the 2:1, 3:1, 5:2 and 7:3 orbital resonances. Asteroids in these resonant orbits get dislodged by repeated gravitational tugs from Jupiter, and hence don't stay there for very long.

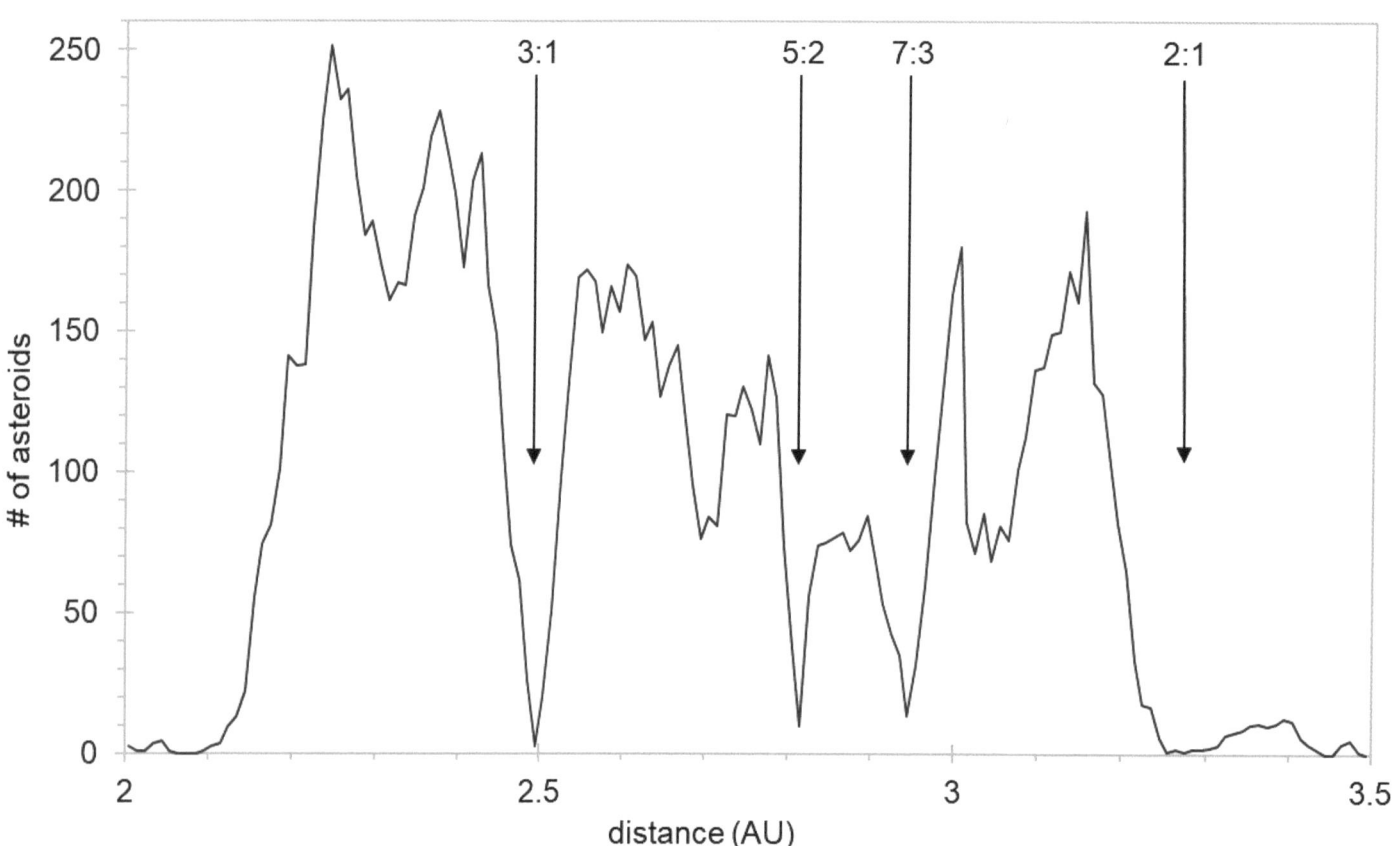

Figure 6D: Distribution of objects in the asteroid belt arranged into 0.01 AU bins.

The Sun's outermost true planet is Neptune and the Kuiper belt represents a sort of transitional wilderness between the planets and the Oort cloud. The main segment of the Kuiper belt lies at ~41.8 AU, near the $n = 7$ value in the Blagg-Richardson formulation. This distance may be partially based on where the material in the planetary nebula thinned out. It could have also been the case that stable orbits for regular planets were not possible there in the early solar system as perturbations from passing stars or galactic tidal influence might have interfered with their formation.

Kuiper belt objects are merely dwarf planetesimals that have eccentric and inclined orbits that are resonant with Neptune. Pluto is a Kuiper belt object that settled into a stable 2:3 orbital resonance with Neptune, a feature it shares with ~200 smaller objects known as Plutinos. Most other Kuiper belt objects can be found in orbits beyond Pluto at distance of up to 50 AU (Stern and Colwell 1997). Based on the distribution of minor planetesimals in our own solar system, it is likely that the concept of orbital resonance plays a key role in determining the structure of debris belts in planetary systems in general.

Exoplanetary systems

The discovery of planets orbiting other stars began when several terrestrial mass planets were detected around the pulsar PSR B1257+12 in 1992 (Wolszczan and Frail 1992). Since then, more than 4000 exoplanetary systems have been discovered, including 875 with more than one confirmed planet (Schneider 2010). In 2013, Erik Petigura claimed that about 1 in 5 sun-like stars might have Earth-sized planets in their habitable zones (Petigura 2013).

Whether or not asteroid belts exist in other planetary systems probably depends on how the planetesimals are arranged in relation to the frost line at the time of a planetary system's formation. Because the same astrophysical principles and mechanisms should apply throughout the universe, many exoplanetary systems probably contain asteroid belts like ours near their frost lines. As in our solar system, the first gas giant planet would be exuding powerful tidal perturbations on rocky planetesimals that may be trying to form on the inward side of the frost line. This will essentially cause any protoplanets in a zone proximally inward of the first gas giant to be ripped apart by tidal forces before even having a chance to form. The general idea of extrasolar asteroid belts occurring in the vicinity of the frost line is in fact supported by observations of dust around 90 nearby stars (Martin and Livio 2012).

A hotter star than our Sun would have a more distant frost line, where the first gas giant may exist further on in a Blagg-Richardson-type sequence. In that case, the planets near the frost line would be more spread out, which would reduce the strength of interplanetary tidal forces in that zone. This means that for a luminous star, the distance between planetary orbits near the frost line may be too wide for the first giant gas planet's tidal forces to significantly affect the next inward rocky planetesimal.

In contrast to this, a dimmer star would have a closer frost line, which would shift the first gas giant to a position closer to the star. This would have the effect of increasing interplanetary tidal forces in the zone around the frost line, allowing proximally inward planetesimals to be more easily ripped apart. This process, however, would be balanced by the fact that stars less luminous than our Sun might have less material in their planetary nebulae and therefore could have less massive gas giants.

Nonetheless, in these lower luminosity planetary systems the increase in interplanetary tidal forces would be the more important factor for the creation of asteroid belts. Considering that >75% of the known stars in our galaxy are red dwarfs less bright than our Sun, asteroid belts may be quite common (Ledrew 2001).

The composition of the solid material in the early stages of a planetary system's formation may also have a measurable influence on whether an asteroid belt can form. Our solar system probably formed from a relatively evenly distributed collection of rocky material, making the creation of an asteroid belt more or less inevitable. On the other hand, planetary systems that form in binary star systems or turbulent interstellar environments may lack the right initial conditions to form protoplanets with large rocky cores past the frost line, preventing gas giants from forming in the first place.

Observational evidence

Observational evidence for Kuiper belt-type debris disks around other stars is extensive, but the evidence for extrasolar asteroid belts is limited (Wyatt 2020). A dust disk around ε Eridani, a nearby star about 10 light years away, was discovered in 1985 (Aumann 1985). Astronomers now believe that the ε Eridani system may contain two asteroid belts, one at ~2 AU and another from ~8 to ~20 AU, in addition to a Kuiper belt-type debris disk at ~100 AU (Backman et al 2008). The total mass orbiting within the inner debris belts of ε Eridani is probably more than 10 Earth masses (Backman et al 2008). Being much higher than the mass of our own asteroid belt, this puts the ε Eridani system's debris disks firmly into the protoplanetary disk category rather than a defined exoplanetary system. Considering that ε Eridani's age is probably less than 800 Myr, a mere fraction of the Sun's age, and that modern astronomers are still debating whether or not the star has actual planets revolving around it, the protoplanetary disk interpretation is quite plausible (Zechmeister et al 2013, Janson et al 2015).

In 2005, a debris ring was detected around HD 69830, an 11-Gyr-old yellow dwarf star located ~41 light years away (Beichman 2005, Tanner et al 2015). HD 69830's debris belt lies at a distance of ~1 AU and carries an estimated mass of 20 times that of the solar system's asteroid belt. Using a Blagg-Richardson-type relation, a 2008 study of the

55 Cancri system, also located ~41 light years away, predicted the occurrence of an asteroid belt at a radius of ~2 AU, but as of now there is no observational evidence for its existence (Bovaird and Lineweaver 2013).

A division between smaller inner planets and larger outer planets on either side of a frost line appears to be present in some known exoplanetary systems. These include Gliese 676, Gliese 876, HD164922, HD219134, Kepler-65, Kepler-90 and μ Arae. Yet a vast number of exoplanetary systems appear not to follow this trend at all, including the well-studied 55 Cancri. The abundance of hot Jupiters, hot Neptunes and super Earths detected around other stars also casts doubt on whether the frost line should be used as a de-facto rule for exoplanetary systems (McNeil and Nelson 2010). However, it should be kept in mind that lower-than-Earth-mass planets are still quite difficult to detect in exoplanetary systems.

Conclusions

The analysis presented here suggests that asteroid belts may be common features in the universe, although instruments refined enough to detect them have not yet been developed. The hypothetical abundance of asteroid belts in the universe is discerned based on two general astrophysical principles observed in our own solar system that can be applied to planetary systems in general. That is, the existence of a frost line and the arrangement of planets following a Blagg-Richardson-type relation.

Asteroid belts may be a regular consequence of planetary system formation. As planetesimals beyond the frost line accumulate H_2, He, H_2O, NH_3 and CH_4 in their atmospheres, they will exhibit strong tidal perturbations on the zone proximally inward of the frost line, preventing large planetesimals from forming there. Due to deterministic mathematical patterns such as the Richardson formulation, this could be a much more common scenario in planetary system formation than previously believed. Asteroid belts may in fact be an inescapable reality for most exoplanetary systems around red dwarfs and small main sequence stars that will eventually be confirmed by future observational evidence.

References

Aumann,H.H.(1985).IRAS observations of matter around nearby stars.*Publications of the Astronomical Society of the Pacific*,97(596),885.

Backman,D.,Marengo,M.,Stapelfeldt,K.,Su,K.,Wilner,D.,Dowell,C.D.,Werner,M.(2008).Epsilon Eridani's planetary debris disk:structure and dynamics based on Spitzer and Caltech Submillimeter Observatory observations.*The Astrophysical Journal*,690(2),1522.

Beichman,C.A.,et al.(2005).An Excess Due to Small Grains around the Nearby K0 V Star HD 69830:Asteroid or Cometary Debris?.*The Astrophysical Journal*.626(2): 1061–1069.

Blagg,M.A.(1913).On a Suggested Substitute for Bode's Law. (Plate 12.).*Monthly Notices of the Royal Astronomical Society*,73(6),414-422.

Bode,J.E.(1772).Deutliche Anleitung zur Kenntniß des gestirnten Himmels:zum gemeinnützigen und beständigen Gebrauch ausgefertiget.*Verf*.

Bovaird,T.,Lineweaver,C.H.(2013).Exoplanet predictions based on the generalized Titius-Bode relation.*Monthly Notices of the Royal Astronomical Society*,435(2),1126-1138.

Gregory,D.(1715).The Elements of Astronomy,Physical and Geometrical.(Vol.1).*J.Nicholson*.

Harwit,M.(1988).Astrophysical concepts(405)*New.York: Springer*.

Hayashi,C.(1981).Structure of the solar nebula, growth and decay of magnetic fields and effects of magnetic and turbulent viscosities on the nebula.*Progress of Theoretical Physics Supplement*,70,35-53.

Herschel,W.(1781).Account of a comet.Philosophical Transactions of the Royal Society of London,(71),492-501.

Huang,C.X.,Bakos,G.Á.(2014).Testing the Titius–Bode law predictions for Kepler multiplanet systems.*Monthly Notices of the Royal Astronomical Society*,442(1),674-681.

Janson,M.,Quanz,S.P.,Carson,J.C.,Thalmann,C.,Lafrenière,D.,Amara,A.(2015).High-contrast imaging with spitzer: Deep observations of Vega,Fomalhaut,and ϵ Eridani.*Astronomy & Astrophysics*,574,A120.

Jewitt,D.,Chizmadia,L.,Grimm,R.,Prialnik,D.(2007).Water in the small bodies of the solar system.*Protostars and Planets V*,1,863-878.

Junfeng,L.,Xiangyuan,Z.,Yun,Z.(2016).Unique dynamics of asteroids.*Mechanics in Engineering*,38(6),603.

Kaufmann,W.J.(1987).HI.Discovering the Universe.

Ledrew,G.(2001).The real starry sky.*Journal of the Royal Astronomical Society of Canada*,95,32.

Le Verrier,U.J.(1846).Recherches sur les mouvements d'Uranus par UJ Le Verrier(Fortsetzung).*Astronomische Nachrichten*,Vol.25,65.

Martin,R.G.,Livio,M.(2012).On the evolution of the snow line in protoplanetary discs.*Monthly Notices of the Royal Astronomical Society:Letters*,425(1),L6-L9.

McNeil,D.S.,Nelson,R.P.(2010).On the formation of hot Neptunes and super-Earths.Monthly Notices of the Royal Astronomical Society,401(3),1691-1708.

Petigura,E.A.,Howard,A.W.,Marcy,G.W.(2013).Prevalence of Earth-size planets orbiting Sun-like stars.*Proceedings of the National Academy of Sciences*,110(48),19273-19278.

Piazzi,G.(1801).Risultati delle osservazioni della nuova stella scoperta il di'I.*gennajo all'Osservatorio reale di Palermo*.

Pitjeva,E.V.,Pitjev,N.P.(2018).Masses of the Main Asteroid Belt and the Kuiper Belt from the motions of planets and spacecraft.*Astronomy Letters*,44,554-566.

Podolak,M.,Zucker,S.(2004).A note on the snow line in protostellar accretion disks.*Meteoritics & Planetary Science*, 39(11),1859-1868.

Poveda,A.,Lara,P.(2008).The exo-planetary system of 55 Cancri and the Titius-Bode law.*Revista mexicana de astronomía y astrofísica*,44(1),243-246.

Richardson,D.E.(1945).Distances of planets from the sun and of satellites from their primaries.*Popular Astronomy*,53,14.

Schneider,J.(2010).Interactive extra-solar planets catalog.*The Extrasolar Planets Encyclopaedia*.

Stern,S.A.,Colwell,J.E.(1997).Collisional erosion in the primordial Edgeworth-Kuiper belt and the generation of the 30-50 AU Kuiper gap.*The Astrophysical Journal*,490(2), 879.

Tanner,A.,Boyajian,T.S.,Von Braun,K.,Kane,S.,Brewer,J.M., Farrington,C.,Schaefer,G.(2015).Stellar parameters for HD 69830, a nearby star with three Neptune mass planets and an asteroid belt.*The Astrophysical Journal*,800(2), 115.

Von Zach,F.X.F.(Ed.).(1806).Monataliche Correspondenz zur Beförderung der Erd-und Himmels-Kunde,herausg.von F. von Zach(Vol.13).*Beckerische Buchhandlung*.

Wolszczan,A.,Frail,D.A.(1992).A planetary system around the millisecond pulsar PSR1257+12.*Nature*.355(6356):145-147.

Wyatt,M.C.(2020).Extrasolar Kuiper belts.The Trans-Neptunian Solar System,351-376.

Zechmeister,M.,Kürster,M.,Endl,M.,Curto,G.L.,Hartman,H., Nilsson,H.,Cochran,W.D.(2013).The planet search programme at the ESO CES and HARPS-IV.The search for Jupiter analogues around solar-like stars.*Astronomy & Astrophysics*,552,A78.

Do large bolide impacts cause antipodal volcanism?

Abstract

When a large bolide object impacts a planet, it sends shock waves through its interior that converge in antipodal regions and promote seismic activity. The kinetic energy and momentum of large impacts has to go somewhere. Antipodal volcanism has actually been the standard interpretation for some surficial features of the Moon and Mercury since the 1970s with little opposition. This process has also likely occurred on Mars, resulting in the enormous yet dormant volcanoes of the Tharsis Bulge, Olympus Mons, Elysium Mons and Alba Mons lying on the opposite side of the planet from the very large craters of Argyre, Hellas and Isidis. Antipodal volcanism may have also occurred on Earth. The Chicxulub impact crater in the Yucatan Peninsula marks the location where a large bolide struck the Earth's surface at the Cretaceous-Tertiary (K-T) boundary, approximately 66 Myr ago. Also associated with the K-T boundary are the Deccan traps, an enormous flood basalt province in Western India. Although the Deccan traps are not exactly antipodal or synchronous with the Chicxulub crater, it's worth considering the possibility that shock waves from the Chicxulub impact could have travelled along pre-existing conduits and plumes through the mantle to enhance seismic activity in the Deccan region.

Introduction

A 1975 paper by Peter Schultz and Donald Gault discussed the seismic effects of large meteorite impacts, and explained the origin of grooved and hilly terrain at the antipodes of large impact basins on the Moon and Mercury (Schultz and Gault 1975). Mercury's largest crater, the 1550-km-wide Caloris Basin, is associated with rugged terrain near the exact antipode of the planet, as illustrated in figure 7A (Schultz and Gault 1975, Murchie et al 2008). Due to Mercury's small size, the idea that large impacts have sent shock waves through the planet's interior to induce disruptive terrain in antipodal regions hasn't received much opposition.

Schultz and Gault claimed that Mars did not show the same type of features. In the late 1970s, James Peterson published several articles about volcanism on Mars, and echoed the claim that Mars' Tharsis region was not associated with any grooved and hilly antipodal terrain (Peterson 1978). In 1994, an article by David Williams presented computer models that challenged these older interpretations. Williams suggested that the Hellas Basin impact event could have produced the dormant antipodal volcano, Alba Mons (Williams and Greeley 1994). Antipodal volcanism was even explored in a 1998 book called the Mars Mystery (Hancock et al 1998). In 2009, Damian Swift used continuum mechanical simulations to model the internal mechanics of antipodal volcanism on the Moon, Mercury, Mars and Earth (Swift and El-Dasher 2009). Other more recent studies and computer simulations have shown a strong case for antipodal volcanism occurring in Mars' past (Citron et al 2018, Zhang et al 2022).

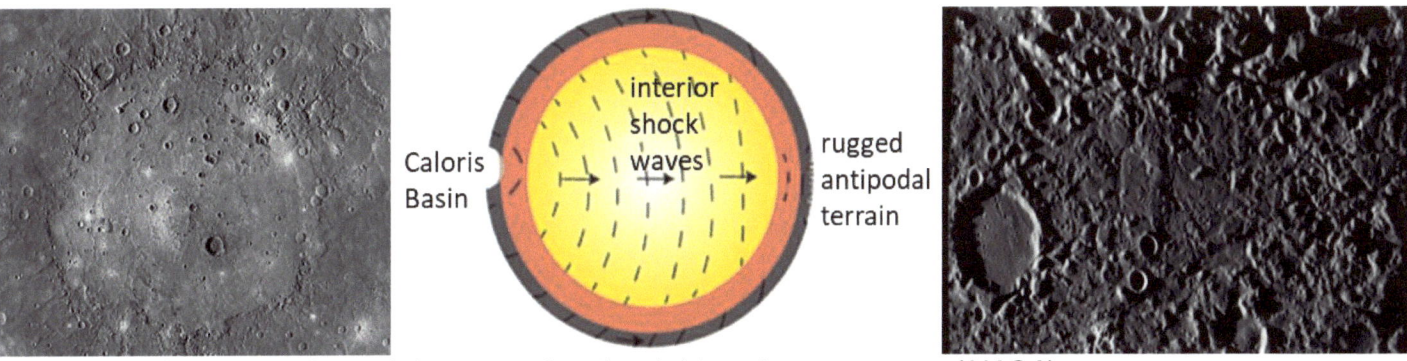

Figure 7A: Caloris basin and the rugged antipodal terrain on mercury (*NASA*).

In the late 1970s, an ancient crater named Chicxulub was discovered in the Yucatan Peninsula. Radiometrically dated to 66.05 Myr old and created by an asteroid ~10 km in diameter, the Chicxulub impact is now thought to have been the primary reason for the extinction of the dinosaurs at the end of the Cretaceous Period (Hildebrand et al 1991, Schulte et al 2010, Renne et al 2013, Renne et al 2018, Desch et al 2021). On the other side of the world from Chicxulub are the Deccan Traps, an enormous flood basalt province in India that is between 69 and 63 Myr old, meaning that the Deccan Traps began several thousand years before the impact event. (Sen 2001, Pande 2002). In 2003, Boris

Ivanov and Jay Melosh calculated that the energy involved in the Chicxulub impact was a few orders of magnitude too small to cause antipodal volcanism (Ivanov and Melosh 2003). A team led by Martin Meschede came to the same conclusion in 2011 after analyzing computer simulations of the Chicxulub impact but added that the energy of the impact could have enhanced volcanism in places where it was already occurring (Meschede et al 2011). Some recent publications also point out that a number of additional undiscovered craters and volcanoes could be associated with the K-T extinction (Schoene et al 2014, Keller et al 2016).

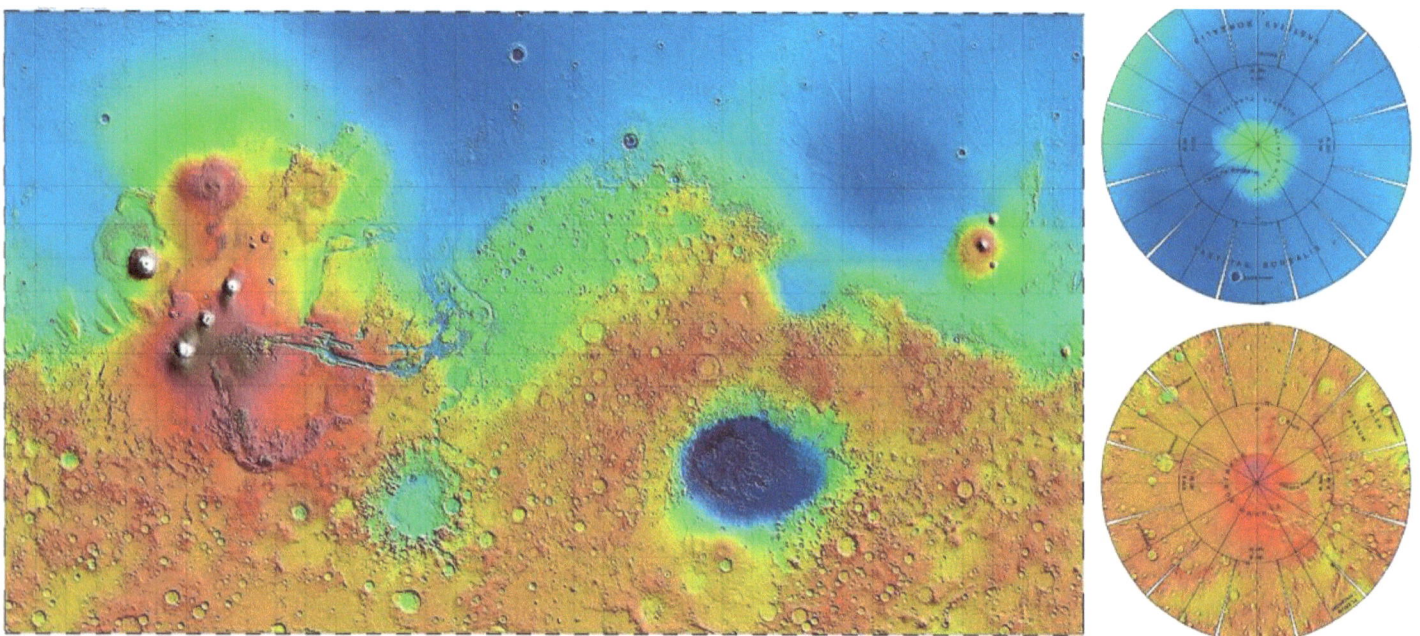

Figure 7B: Map of Martian surface features obtained by the Mars Orbiter Laser Altimeter aboard the Mars Global Surveyor Spacecraft (*NASA*).

Mars

Mars has the largest volcanoes in the solar system, including the 26-km-high Olympus Mons, which is many times higher than Mount Everest with a base as wide as Poland (Comins 2012). Approximately 1200 km southeast of Olympus Mons is the Tharsis Bulge, a set of three monstrous volcanoes resting on an enormous platform 500 km across and 7 km high (Carr 2007). About 2300 km to the west of Olympus Mons lies Elysium Mons, a 240-km-wide volcano with an altitude of 14 km (Plescia 2004). About 1000 km to the northeast of Olympus Mons is Alba Mons, which is actually Mars' widest volcano with lava flows extending up to 1350 km from its center (Cattermole 1990). All of

these volcanoes are completely dormant and are clustered on one hemisphere of the planet, as you can see clearly in figure 7B (Yin 2012).

The surface of Mars also contains many of the solar system's largest craters, which all exist in the opposite hemisphere from where all of Mars' large volcanoes are found. The largest crater is the 2300-km-wide and 7-km-deep Hellas Basin, which is nearly antipodal to Alba Mons (Peterson 1978, Smith et al 1999). The terrain surrounding the Hellas Basin is littered with smaller craters and rugged terrain for thousands of kilometers. The second largest crater is Argyre, which is 1800 km wide, 5.2 km deep and nearly antipodal to Elysium Mons (Williams and Greeley 1994). Another large crater named Isidis is 1500 km in diameter (Ritzer and

Hauck 2009). The combined antipode of Hellas and Isidis lies near the center of the Tharsis bulge.

The two hemispheres of the planet, one with large dormant volcanoes and the other heavily cratered, are divided by the line of Martian dichotomy. In figure 7B, the line of dichotomy is an imaginary boundary running roughly between the Northern Martian lowlands in blue and the Southern Martian highlands in orange. One striking surface feature that exists near the line of dichotomy is the Valles Marineris, a gigantic trench-like system more than 4000 km long and up to 11 km deep (Peulvast et al 2001). The surface close to the line of dichotomy also contains a lot of high mesas and knobs surrounded by flat-floored valleys and fretted terrain (Greeley and Guest 1987). The line of Martian dichotomy is also intersected by a large number of outflow channels (Watters et al 2007). Based on all of this evidence, some planetary scientists have proposed that the line of dichotomy represents the result of a single large impact event that struck Mars' North Polar region before the Hellas Basin formed (Andrews-Hanna et al 2008).

Further evidence for antipodal volcanism in Mars' past is discerned from photos of the Martian landscape obtained by the Viking landers of the late 1970s and the Martian rovers of the 21st century. Although both Viking probes landed thousands of kilometers away from the Tharsis volcanoes and Mars' largest craters, the surface around their landing site was covered in countless basaltic boulders lying on top of red soil, as you can see in figure 7C. The rocks resemble either volcanic ejecta or meteoric debris, and not rocks that naturally formed on the Martian surface. They also don't look extensively worn by the Martian weather.

Figure 7C: The Viking 2 lander site (*NASA*)

Outside of Mars' North polar ice cap, no water can be observed anywhere on its surface (Carr 1996). However, it is possible that Mars could contain a fair amount of subsurface water in liquid or solid form that could have moved quickly along the surface in the wake of a large impact event (Bibring et al 2005). Large outflow channels on the Martian surface, including the 1600-km-long Kasei Valles, bear resemblance to patterns left behind after flash floods on Earth (Coleman and Lindberg 2013). Features resembling large terminal moraines have also been identified near the line of Martian dichotomy, which some scientists have interpreted as evidence for ancient Martian tsunamis (Costard et al 2017).

It's not known exactly when the Hellas impact occurred but a crater of its size could have caused a planet-wide extinction event, if Mars ever had life. It is more likely, however, that such an enormous impact would have occurred in the days of the early solar system as there would have been more asteroids and planetoids in Mars' vicinity then. Erosional processes on Mars operate slowly because water doesn't normally flow on the surface and the atmosphere is thin. Therefore, even if Mars' surficial features were created early on in the planet's history, they may have avoided significant alteration over the last 4.5 Gyr.

Although other impacts could have triggered antipodal volcanism in Mars' past, the Hellas and Isidis impacts were likely the most recent major impact events. This is because the Hellas and Isidis craters have far fewer craters imprinted on top of them compared to the surrounding terrain. In contrast, the Argyre Basin has two other large craters imprinted on top of it, so it is probably older than the Hellas Basin.

Cretaceous-Tertiary Extinction

The Chicxulub impact ~66 Myr ago released so much energy that it vaporized the ocean around it, blanketed the entire earth in a layer of iridium rich ash, and set off forest fires thousands of kilometers away. This made it more destructive than if all the nuclear weapons in the world were gathered into one place and detonated simultaneously. The Chicxulub impact site is surrounded by a large number of cenotes, pit-like holes with extensive underwater caverns.

On the other side of the world from Chicxulub are the Deccan traps, a set of volcanic flood basalts near Mumbai that were very active in the years between ~69 Myr and ~63 Myr, around the Cretaceous-Tertiary (K-T) boundary (Sen 2001, Pande 2002). They are among the most extensive lava flows in Earth's history, at more than 2 km thick and covering ~500,000 km^2 (Singh and Gupta 1994, Renne et al 2015). Although the Deccan traps are slightly older than the Chicxulub crater, the main phase of the volcanic outpourings overlaps the Chicxulub event (Pande 2002, Desch et al 2021). A 2015 paper by Paul Renne claimed that the Deccan traps may have been at least partially induced by the Chicxulub impact, due to their nearly antipodal positions, as you can see in figure 7D (Renne et al 2015). This view is contrasted greatly by Martin Meschede, who has repeatedly pointed out that the Deccan traps are not exactly antipodal to Chicxulub (Meschede et al 2011).

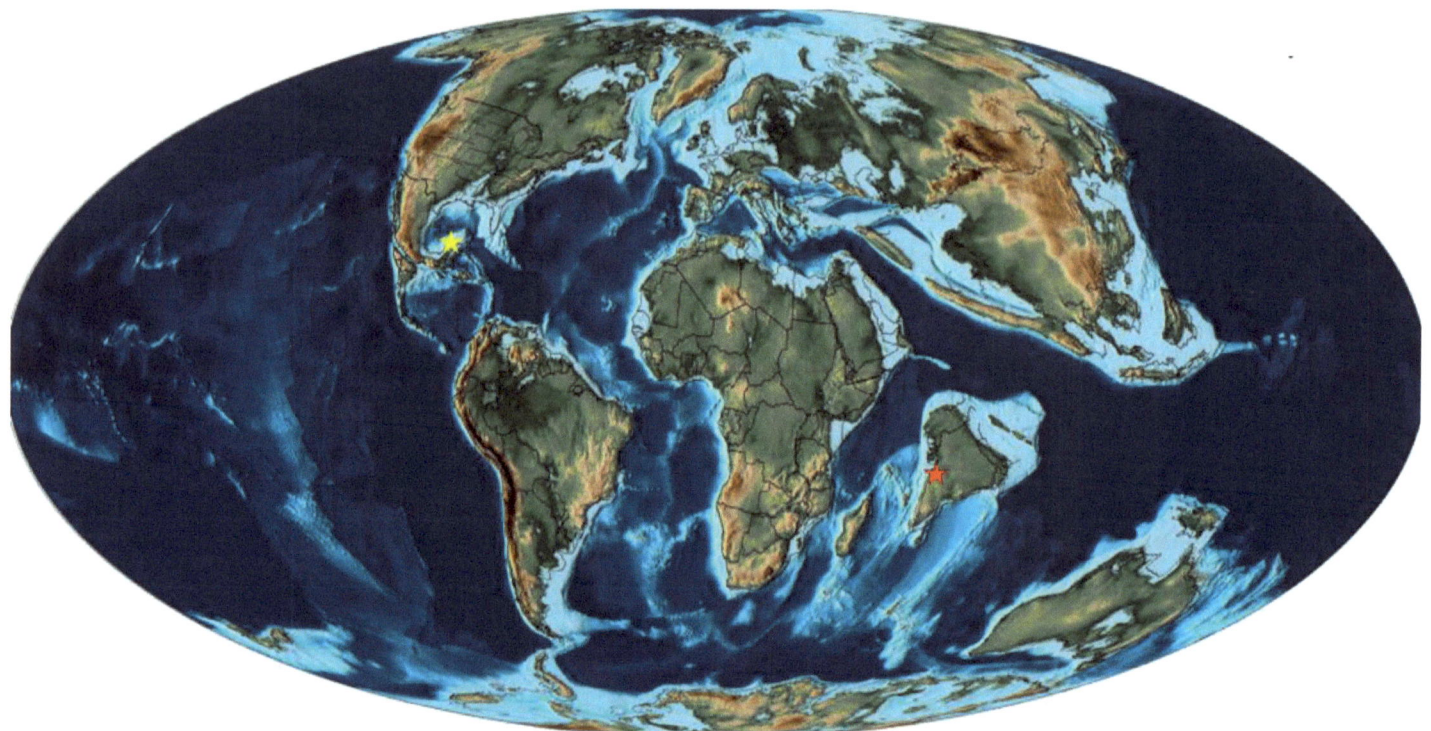

Figure 7D: Map of the Earth near the K-T boundary, 66 Myr ago. The yellow star in Mexico shows the location of the Chicxulub impact crater. The red star in India indicates the central part of the Deccan Traps (Scotese 2014).

Discussion

The Deccan traps and the Chicxulub crater are not exactly antipodal, but the truth is they may not have to be to show a connection with antipodal volcanism. It is in fact possible that the Deccan traps formed earlier than the Chicxulub crater and that they were enhanced for a brief interval by the shock waves of the impact. In other words, since a path through the mantle and crust had already been established, the energy from the shock waves could have been funnelled toward the Deccan traps.

Meteorites of various sizes have bombarded the surfaces of rocky bodies of the inner solar system for billions of years. Most of these meteorites struck in the early days of the solar system and the motion of Earth's tectonic plates may be responsible for erasing a lot of the original craters. That being said, it's worth considering that Chicxulub may not have been the only impact site responsible for the Cretaceous-Tertiary mass extinction and the Deccan traps might not have been the only antipodal volcanic eruption. There is no reason to doubt the possibility that several other craters or extinct volcanoes associated with the K-T extinction may exist, and that some may have even been subducted under the crust.

If the Deccan traps could in fact be the result of antipodal volcanism, then many other mass extinctions in Earth's past could have been based on similar processes. The fact that large bolide impacts have a ~70% chance of hitting an ocean means that direct evidence for impact-triggered extinctions may not be easy to come by. In other words, although a bolide impact in the ocean could have similar disastrous effects for life on Earth recorded in the fossil record, they would be much less visible in the geological record. Therefore, the lack of a direct correlation between large impact craters and associated antipodal volcanoes in Earth history should not cause one to completely discount the idea of antipodal volcanism occurring in Earth's past.

Conclusions

Here we have briefly explored evidence for impact-induced antipodal volcanism on Mercury, Mars and our own planet. Antipodal volcanism has been the favoured explanation for prominent features on the surface of Mercury since the mid 1970s but its application to features on Mars and Earth was originally overlooked. A growing number of scientists are now thinking that antipodal volcanism may have had an important role in creating the large yet dormant volcanoes on Mars and are questioning previous claims that antipodal volcanism on Mars is impossible. The fact that Mars is not tectonically active at all suggests that impact-induced antipodal volcanism may in fact be the only process that could have produced such enormous volcanic

structures. Many scientists are also now entertaining the possibility that antipodal volcanism may have played a role in enhancing the lava flows of the Deccan traps during the Cretaceous-Tertiary extinction, even though they are not entirely antipodal to the Chicxulub crater.

The energy and momentum of large bolide impacts has to go somewhere. It is irresponsible to think that all of the momentum of a large impact is dissipated completely in the terrain around the impact site. Although many scientists of the past seriously underestimated the ability for the momentum of a large impact to be transferred through planetary interiors, evidence for this process occurring on several bodies of the inner solar system is becoming increasingly difficult to ignore. Antipodal volcanism might not just be a possible explanation for a multitude of features on the surface of the Moon, Mercury, Mars and Earth, but a necessary process for describing the surfaces of all rocky planets.

Acknowledgments

Christopher Scotese is thanked for the use of his paleogeographic map in this article. Paul McNeil is thanked for reviewing this article and providing insightful comments.

References

Andrews-Hanna,J.C.,Zuber,M.T.,Banerdt,W.B.(2008).The Borealis basin and the origin of the martian crustal dichotomy.*Nature*,*453*(7199),1212-1215.

Bibring,J.P.,Langevin,Y.,Gendrin,A.,Gondet,B.,Poulet,F., Berthé,M.(2005).Mars surface diversity as revealed by the OMEGA/Mars Express observations.*Science*,*307*(5715), 1576-1581.

Carr,M.H.(1996).Water on mars.*New York:Oxford University Press*.

Carr,M.H.(2007).The surface of Mars (Vol.6).*Cambridge University Press*.

Cattermole,P.(1990).Volcanic flow development at Alba Patera,Mars.*Icarus*,*83*(2),453-493.

Citron,R.I.,Manga,M.,Tan,E.(2018).A hybrid origin of the Martian crustal dichotomy:Degree-1 convection antipodal to a giant impact.*Earth and Planetary Science Letters*,*491*,58-66.

Coleman,N.M.,Lindberg,S.(2013,March).New Insights About Cataracts(Dry Falls)on the Floor of Kasei Valles,Mars. In *44th Annual Lunar and Planetary Science Conference* (No.1719,1148).

Comins,N.F.(2012).Discovering the essential universe.*Macmillan*.

Costard,F.,Séjourné,A.,Kelfoun,K.,Clifford,S.,Lavigne,F.,Di Pietro,I.,Bouley,S.(2017).Modeling tsunami propagation and the emplacement of thumbprint terrain in an early Mars ocean.*Journal of Geophysical Research:Planets*, *122*(3), 633-649.

Desch,S.,Jackson,A.,Noviello,J.,Anbar,A.(2021).The Chicxulub impactor:comet or asteroid?*Astronomy & Geophysics*,*62*(3),3-34.

Greeley,R.,Guest,J.(1987).Geologic map of the eastern equatorial region of Mars(Vol.1)*US Geological Survey*.The Survey.

Hancock,G.,Bauval,R.,Grigsby,J.(1998).The Mars Mystery:A Tale of the End of Two Worlds.*Michael Joseph Ltd*.

Hildebrand,A.R.,Penfield,G.T.,Kring,D.A.,Pilkington,M.,Camargo,Z,A.,Jacobsen,S.B.,Boynton,W.V.(1991).Chicxulub crater:a possible Cretaceous/Tertiary boundary impact crater on the Yucatan Peninsula,Mexico.*Geology*,*19*(9), 867-871.

Ivanov,B.A.,Melosh,H.J.(2003).Impacts do not initiate volcanic eruptions:Eruptions close to the crater.*Geology*, *31*(10),869-872.

Keller,G.,Punekar,J.,Mateo,P.(2016).Upheavals during the late Maastrichtian: Volcanism, climate and faunal events preceding the end-Cretaceous mass extinction.*Palaeogeography,Palaeoclimatology,Palaeoecology*,*441*,137-151.

Meschede,M.A.,Myhrvold,C.L.,Tromp,J.(2011).Antipodal focusing of seismic waves due to large meteorite impacts on Earth.*Geophysical Journal International*,*187*(1),529-537.

Murchie,S.L.,Watters,T.R.,Robinson,M.S.,Head,J.W.,Strom, R.G.,Chapman,C.R.,Blewett,D.T.(2008).Geology of the Caloris basin,Mercury:A view from MESSENGER.*Science*,*321*(5885),73-76.

Pande,K.(2002).Age and duration of the Deccan Traps,India: A review of radiometric and paleomagnetic constraints.*J. Earth Syst.Sci*,*111*(2),115.

Peterson,J.E.(1978, March).Antipodal effects of major basin-forming impacts on Mars.In *Lunar and Planetary Science Conference*(Vol. 9,885-886).

Peulvast,J.P.,Mège,D.,Chiciak,J.,Costard,F.,Masson,P.L. (2001).Morphology,evolution and tectonics of Valles Marineris wallslopes(Mars).*Geomorphology*,*37*(3-4),329-352.

Plescia,J.B.(2004).Morphometric properties of Martian volcanoes.*Journal of Geophysical Research*.109(E3):E03003.

Renne,P.R.,Deino,A.L.,Hilgen,F.J.,Kuiper,K.F.,Mark,D.F., Mitchell III,W.S.,Smit,J.(2013).Time scales of critical events around the Cretaceous-Paleogene boundary.*Science*,*339*(6120),684-687.

Renne,P.R.,Sprain,C.J.,Richards,M.A.,Self,S.,Vanderkluysen, L.,Pande,K.(2015).State shift in Deccan volcanism at the Cretaceous-Paleogene boundary,possibly induced by impact.*Science*,*350*(6256),76-78.

Renne,P.R.,Arenillas,I.,Arz,J.A.,Vajda,V.,Gilabert,V., Bermúdez,H.D.(2018).Multi-proxy record of the Chicxulub impact at the Cretaceous-Paleogene boundary from Gorgonilla Island,Colombia.*Geology*,*46*(6),547-550.

Ritzer,J.A.,Hauck II,S.A.(2009).Lithospheric structure and tectonics at Isidis Planitia,Mars.*Icarus*,*201*(2),528-539.

Schoene,B.,Samperton,K.M.,Eddy,M.P.,Keller,G.,Adatte,T., Bowring,S.A.,Khadri,S.F.R.,Gertsch,B.(2014).U-Pb geochronology of the Deccan Traps and relation to the end-Cretaceous mass extinction.*Science*.347(6218):182–184.

Schulte,P.,Alegret,L.,Arenillas,I.,Arz,J.A.,Barton,P.J.,Bown, P.R.,Willumsen,P.S.(2010).The Chicxulub asteroid impact and mass extinction at the Cretaceous-Paleogene boundary.*Science*,*327*(5970),1214-1218.

Schultz,P.H.,Gault,D.E.(1975).Seismic effects from major basin formations on the Moon and Mercury.*The Moon*,*12*(2), 159-177.

Scotese,C.(2014).Atlas of Late Cretaceous Paleogeographic Maps,PALEOMAP Atlas for ArcGIS,volume 2,The Cretaceous,Maps 16-22,Mollweide Projection, PALEOMAP Project,Evanston,IL.

Sen,G.(2001).Generation of Deccan trap magmas.*Journal of Earth System Science*,*110*(4),409-431.

Singh,R.N.,Gupta,K.R.(1994).Workshop yields new insight into volcanism at Deccan Traps, India.*Eos*.75(31):356.

Smith,D.E.,Zuber,M.T.,Solomon,S.C.,Phillips,R.J.,Head,J.W., Garvin,J.B.,Duxbury,T.C.(1999).The global topography of Mars and implications for surface evolution.*Science*, *284*(5419),1495-1503.

Swift,D.,El-Dasher,B.(2009,September).Planetary Structures And Simulations Of Large-scale Impacts On Mars.In *AAS/ Division for Planetary Sciences Meeting Abstracts#41*(Vol. 41,58-03).

Watters,T.R.,McGovern,P.J.,Irwin III,R.P.(2007).Hemispheres apart:The crustal dichotomy on Mars.*Annu.Rev.Earth Planet.Sci.*,*35*,621-652.

Williams,D.A.,Greeley,R.(1994).Assessment of antipodal-impact terrains on Mars.*Icarus*,*110*(2),196-202.

Yin,A.(2012).Structural analysis of the Valles Marineris fault zone:Possible evidence for large-scale strike-slip faulting on Mars.*Lithosphere*,*4*(4),286-330.

Zhang,L.,Zhang,J.,Mitchell,R.N.(2022).Dichotomy in crustal melting on early Mars inferred from antipodal effect.*The Innovation*,*3*(5),100280.

South Pole		North Pole
0°, 0°	14.7 kyr ago –	0°, 0°
70°S, 130°E	130 – 14.7 kyr ago	70°N, 50°W
77°S, 169°W	243 – 130 kyr ago	77°N, 11°E
85°S, 175°W	337 – 243 kyr ago	85°N, 5°E
75°S, 110°E	424 – 337 kyr ago	75°N, 70°W
75°S, 105°E	– 424 kyr ago	75°N, 75°W

Date

X 0
X 1
X 2
X 3
X 4
X 5

The North Pole's path in the Late Pleistocene

Abstract

During the Last Glacial Maximum (LGM) ~20 kyr ago, continental ice extended all the way down to Wisconsin at around 40°N latitude. At the same time, large parts of Alaska and Eastern Siberia were unglaciated. This asymmetric distribution of the ice sheets suggests that the North Pole may have been located at a position in Central Greenland during the LGM. It could have then migrated to its current location during the rapid melting of continental ice sheets around the Pleistocene-Holocene transition. The implications of an alternate LGM North Pole location as well as a number of other previous polar migrations during the Late Pleistocene are explored in this study. A migration of Earth's rotational axis could have resulted from changes in Earth's center of mass caused by a redistribution of water on Earth's surface as continental ice sheets melted during the Pleistocene-Holocene transition. Calculations presented here suggest that changes in the distribution of mass on Earth's surface during the Pleistocene-Holocene transition could have moved the position of the North Pole by ~23 km. This is a small fraction of the distance from Central Greenland to the North Pole's current location, so in order for this scenario to be true there has to be some additional process that would have amplified the change in pole position.

Introduction

20 kyr ago, during the Last Glacial Maximum (LGM), continental ice sheets up to ~1 km thick covered most of Canada and extended all the way to the Ohio valley and Manhattan (Carlson et al 2018, Pico et al 2018, Yu et al 2019). Ice sheets covered much of Northern Europe during this time but much of Alaska and Eastern Siberia was ice-free, as shown in Figure 8A. The average global temperature during the LGM was ~6°C colder than today and so much of the world's water was tied up in continental ice that the sea level was lower by ~125 meters (Poore et al 2000, Tierney et al 2020). This caused Australia, New Guinea and Tasmania to be connected, and many Indonesian islands to be part of a larger, contiguous version of Southeast Asia known as Sundaland (Bird et al 2005). North America was attached to Siberia by a land bridge known as Beringia, which had light annual snowfall and supported a steppe grassland environment interspersed with trees such as larch, spruce, birch, poplar and alder (Hulten 1937, Hopkins 1967, Voous 1973, Hoffecker and Elias 2007). Herds of mammoth, bison and horses roamed this ancient land (Zazula 2003).

A question must now be asked: if the world was colder during the LGM, why weren't large ice sheets covering Beringia? The usual explanation given by scientists is that different climate patterns during the LGM stopped large amounts of precipitation from falling on Beringia, which prevented snow and ice from accumulating there (Felzer 2001). This explanation has always been problematic, however, as if Beringia had inadequate precipitation, then it couldn't have supported grasses, trees and herds of mammoths that would have needed freshwater to drink. It's also worth noting that although today's high Arctic region is identified as a polar desert, much of Ellesmere Island is covered in snow all year round. Therefore, even though precipitation is light in the high Arctic, the snow doesn't melt quickly. Considering this, there is no good reason to believe that snow and ice would have been prevented from accumulating in Beringia during the LGM. Placing the North Pole in Central Greenland during the LGM rectifies this conundrum. However, before going into further detail on the potential migration of the North Pole at the Pleistocene-Holocene transition, it is important to explain some essential background information about the Pleistocene Ice Ages.

← **Figure 8A**: Earth's surface during the LGM. Overlaying the map is a set of X's to mark previous positions of the North and South Poles inferred from the changing distribution of ice sheets in the Late Pleistocene.

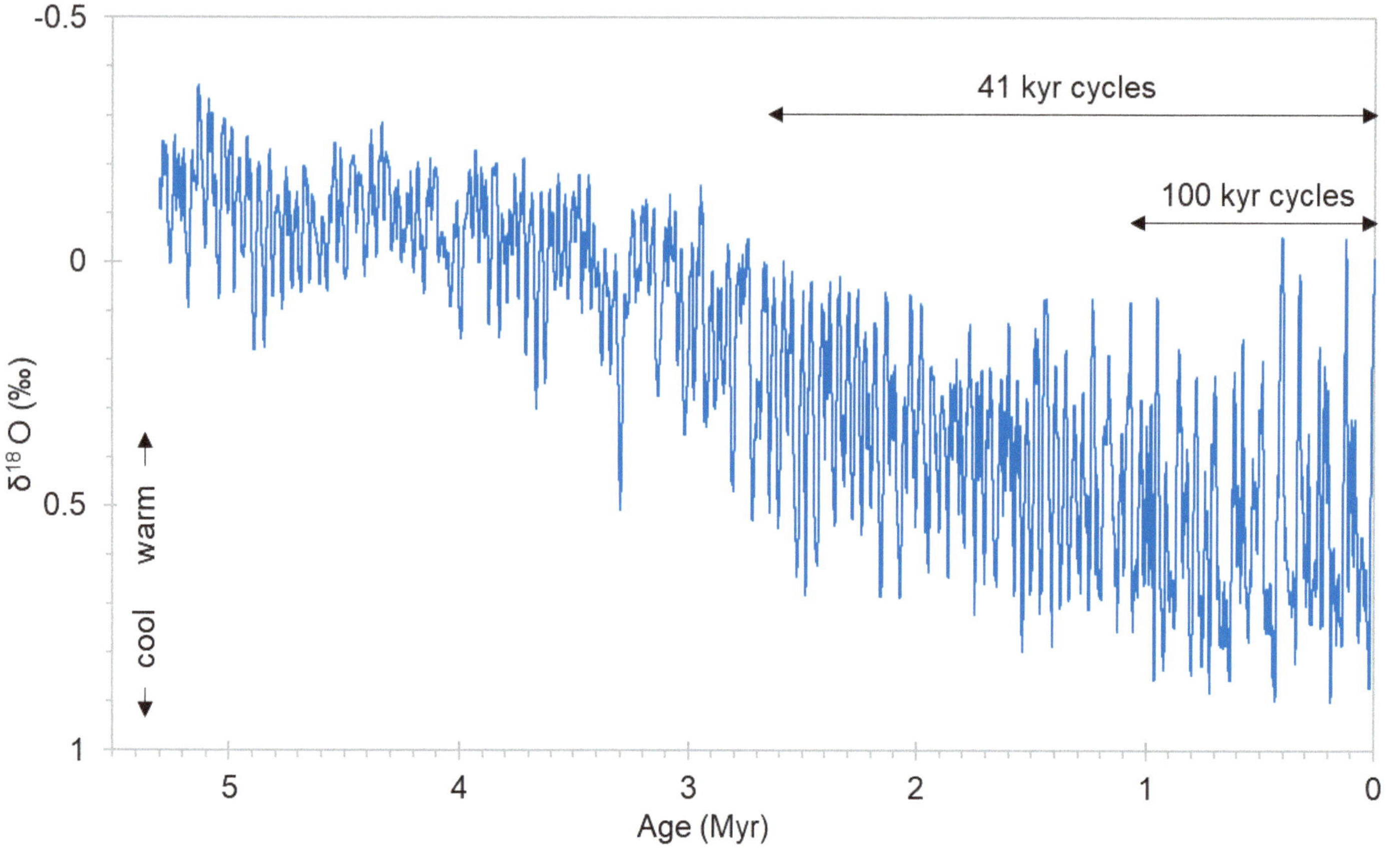

Figure 8B: $\delta^{18}O$ paleotemperature data over the last 5.5 Myr from ocean sediment (DeBoer et al 2013). Regular large-scale oscillations of ~100 kyr have occurred over the last ~1.0 Myr. Smaller amplitude oscillations of ~41 kyr have occurred over the last ~2.7 Myr.

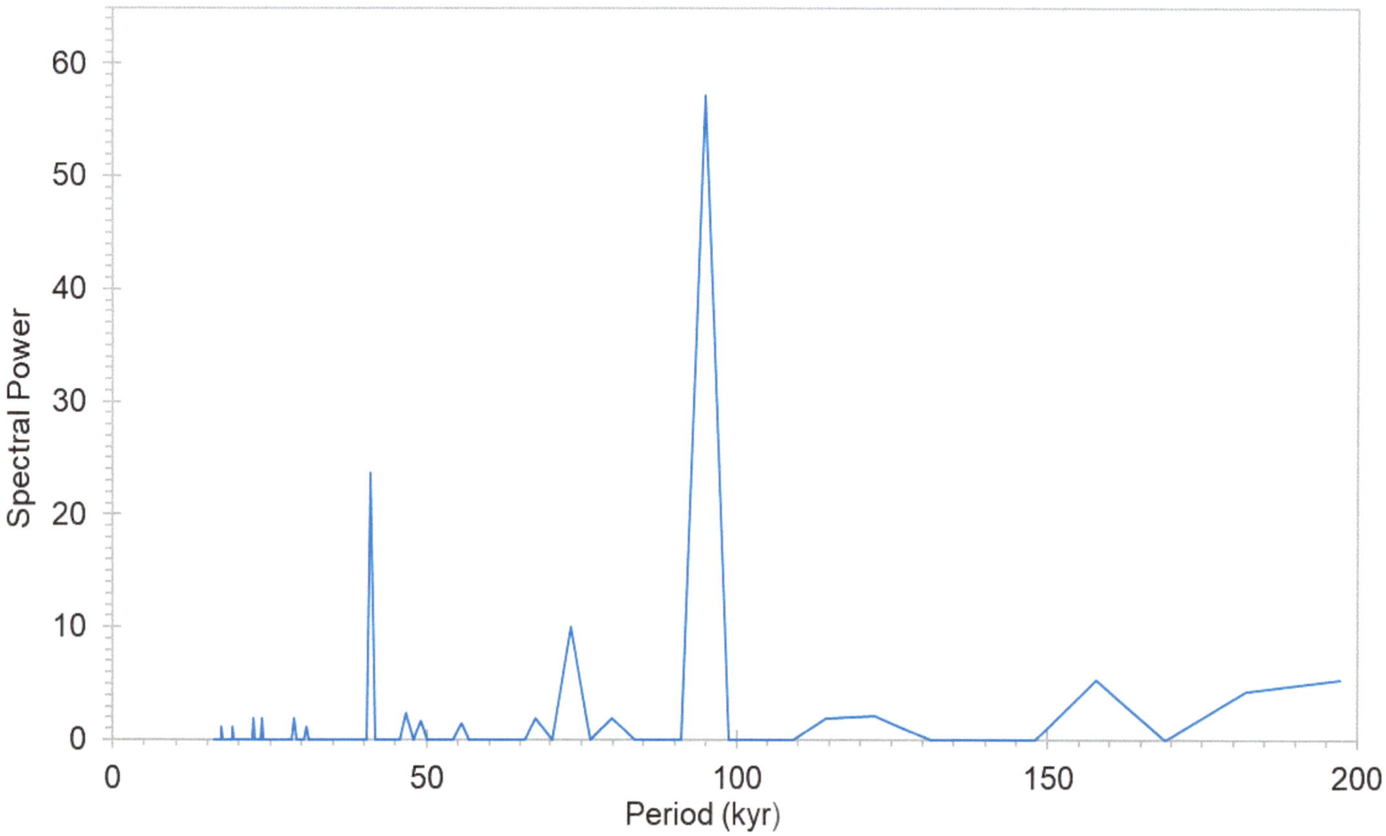

Figure 8C: Spectral analysis of $\delta^{18}O$ ocean sediment data from the last 900 kyr (Schulz and Zeeve 2006). Prominent peaks at 94 kyr and 41 kyr are thought to be related to Earth's eccentricity and obliquity cycles.

The Pleistocene Ice Ages

In 1824, Jens Esmark suggested that Earth has gone through a series of glacial and interglacial episodes over hundreds of millennia, and that these glacial-interglacial oscillations were based on changes in Earth's orbit (Esmark 1824). In 1840, Louis Agassiz published extensive fieldwork, showing clear evidence for vast changes in Earth's ice cover and repeated glacial advances in Europe and North America (Agassiz and Bettannier 1840). By the 1870s, the existence of ice ages in Earth's past was accepted by most scientists. In 1885, James Croll proposed a mathematical explanation that linked the coming and going of glacial and interglacial periods to changes in Earth's orbit over thousands of years, which further popularized the theory of ice ages (Croll 1885). In the 1920s, Milutin Milankovitch presented updated calculations that related changes in Earth's orbit to high-latitude solar radiation (Milankovitch 1941). Since the Northern Hemisphere contains more continental landmass to build ice sheets on, it is widely believed that the amount of sunlight received north of 60°N latitude is the most important factor for the growth and retreat of ice sheets.

Milankovitch identified three cyclic changes in Earth's orbit that affect the annual sunlight received at high latitudes. The first is Earth's eccentricity cycle, which causes its orbit to vary from nearly circular to slightly elliptical over a period of ~100 kyr. Eccentricity variation is caused by the gravitational influence of other planets, primarily Jupiter, Venus and Saturn, which give Earth periodic gravitational tugs. The eccentricity cycle affects the total annual sunlight Earth receives according to Kepler's second law of planetary motion, which states that a planet sweeps out equal areas of its elliptical orbit in equal times. A more elliptical orbit reduces total annual sunlight as Earth spends more time in the further part of the ellipse, where it travels slower, and less time in the closer parts of the ellipse, where it moves faster. Earth's eccentricity reached a maximum ~12 kyr ago and has been slowly decreasing to its current value of 0.0167 (Laskar et al 2011). The ~100 kyr eccentricity cycle has been the dominant driver of glacial-interglacial oscillations over the last 1.0 Myr or so, as seen in figure 8B's deep-sea sediment record. A spectral analysis of the deep-sea $\delta^{18}O$ record over the last 900 kyr is shown in figure 8C.

The second cycle Milankovitch described relates to changes in Earth's axial tilt, otherwise known as obliquity. All planets in the solar system have some degree of obliquity, with Uranus leading the pack at 98°. Earth's obliquity varies from 22.1° to 24.5° with respect to the ecliptic over a regular ~41 kyr cycle. Earth's axis is currently tilted 23.4°, and it has been declining since the last obliquity maximum ~10.7 kyr ago (Buis 2020). Earth's obliquity is quite stable as the gravitational influence of the Moon prevents long term axial drift (Ward 1982). The obliquity cycle is caused by the gravitational influence of other planets, who exert periodic torques on Earth's equatorial bulge. More axial tilt causes more extreme seasons, leading to more melting of high-latitude ice during summer. The obliquity cycle has had a major effect on the growth and retreat of ice sheets for at least the last ~2.7 Myr. According to the $\delta^{18}O$ record, the influence of the ~41 kyr cycle on Earth's climate is about half as strong as the influence of the ~100 kyr eccentricity cycle.

The third periodicity identified by Milankovitch is the precessional cycle, often described as a wobble in Earth's axis. The precessional cycle results from the continuous gravitational influence of the Sun and the Moon on Earth's equatorial bulge. The sidereal period of Earth's precession is ~26 kyr, over which the North Celestial Pole traces out a circle (Seidelmann 1992). The current North star is α Ursae Minoris (Polaris), but 2000 years ago it would have been β Ursae Minoris (Kochab) and 12 kyr ago it would have been α Lyrae (Vega). The current precessional layout places Earth's perihelion, its closest approach to the Sun, on January 2nd when the Northern Hemisphere points directly away from the Sun. Due to precession, this would have been reversed ~12 kyr ago as the Northern Hemisphere would have pointed directly towards the Sun at perihelion. The maximal pulse of sunlight on the Northern Hemisphere's continental ice sheets would have caused large amounts of ice to melt at enhanced rates. Note that the effect of Earth's precession on climate is modulated by eccentricity, as it also increases solar radiation during perihelia.

The existence of the three Milankovitch cycles was confirmed in the 1970s by $\delta^{18}O$ records from deep-sea sediment cores (Hays et al 1976). The prominent periodicities in figure 8B, ~94 kyr and ~41 kyr, are close to the periods of the eccentricity and obliquity cycles. Although the deep-sea $\delta^{18}O$ record led to the widespread acceptance of Milankovitch's theory, it revealed many surprises.

Croll and Milankovitch believed that precession would have the largest effect on total annual sunlight, yet the dominant periodicity is not ~26 kyr. This can be reconciled by the idea that the ~100 kyr oscillations are based on eccentricity modulation of the precessional cycle. As every fourth precessional maximum occurs during an orbit of higher eccentricity, the most extreme melting occurs every ~100 kyr during the overlap of eccentricity and precessional maxima. If Earth's orbit was a perfect circle, the precessional cycle would have markedly less influence on climate.

More surprises came when looking further back into the temperature record. Figure 8B shows $\delta^{18}O$ over the last 5.5 Myr and shows that the ~100 kyr cycle has only had a major influence on temperature over the last ~1.0 Myr, where it has dominated long-term climate oscillations (Pisias and Moore 1981). In the last 1.0 Myr, the ~41 kyr obliquity cycle has been secondary to the ~100 kyr cycle, but from ~2.7 Myr ago to ~1.0 Myr ago it took a turn as the main periodicity as the ~100 kyr cycle was absent. Before ~2.7 Myr ago, the ~41 kyr cycle was absent and Earth's climate followed lower-amplitude precessional oscillations of ~19 kyr and ~23 kyr (Tiedemann et al 1994). Since Milkankovitch cycles have been affecting Earth's orbit perpetually over the last 5.5 Myr, what caused the ~41 kyr cycle to suddenly switch on ~2.7 Myr ago and the ~100 kyr cycle to do the same ~1.0 Myr ago?

In trying to understand this strange behaviour of Milankovitch cycles' effect on the climate record, it's useful to note the underlying trend that Earth's temperature has gradually cooled over the last 5.5 Myr. This larger-scale cooling trend is unrelated to Milankovitch forcing. In fact, Earth has been gradually cooling since ~100 Myr ago, when no ice existed on Earth's surface. This long-term cooling trend eventually led to the formation of an Antarctic ice cap ~34 Myr ago, and by ~2.9 Myr ago year-round ice began amassing on Greenland. The initiation of the Greenland ice cap roughly coincides with when the ~41 kyr cycle manifested in the climate record (Bartoli et al 2005, Liu et al 2009). Although obliquity had been changing the whole time, its effect became more important once Greenland had a large enough collection of ice that could advance and retreat based on the changing tilt. This in turn led to larger-scale oscillations in climate. Positive feedback loops would have amplified climate oscillations once the Northern Hemisphere

had a lot of ice. Once year-round ice began covering much of Greenland, the Northern Hemisphere's albedo increased, causing more light to be reflected from Earth's surface, leading to further cooling. However, the maximal amount of sunlight received during obliquity maxima was enough to reverse this trend and cause deglaciation.

The ~100 kyr cycle replaced the ~41 kyr as the dominant oscillation period ~1.0 Myr ago in an event that probably marked a time when a threshold was crossed that allowed sea ice to collect in the Arctic Ocean. After breaching this threshold, sea ice amassed but remained vulnerable to rapid ablation during times of maximal Northern Hemisphere sunlight such as high-eccentricity perihelia. This situation would have allowed sea ice to slowly collect during glacial phases but melt quickly at the start of interglacial phases.

Figure 8D zooms in to the last 450 kyr. During that period, temperature and sea level generally followed a sawtooth pattern of lengthy glacial phases and shorter interglacial phases, primarily controlled by the eccentricity and obliquity cycles. This repeating pattern of rapid warmings and gradual coolings is based on thermodynamic properties of ice, as it tends to melt more quickly than it accumulates. The rapid melting at the end of glacial periods is amplified by positive feedback mechanisms including ice-albedo feedback and changes in ocean currents. Warming phases at the end of glacial periods could also be intensified if accompanied by a relocation of the North and South poles.

Figure 8E zooms in even further to the last 20 kyr and uses data from Greenland and Antarctic ice cores, which give higher resolution records than ocean sediment. We are currently in an interglacial phase that corresponds with the Holocene. The LGM lasted between ~22 and ~18 kyr ago and ended abruptly in two pulses: the Bolling-Allerod warming event ~14.7 kyr ago and the Pleistocene-Holocene transition ~11.7 kyr ago (Rasmussen et al 2006, Walker et al 2009, Tierney et al 2020). The Bolling-Allerod event was coupled with a sea level rise of 16-25 m within half a century known as meltwater pulse 1A (Lavoie et al 2002, Rohde 2005, Cronin 2012). The rate of melting slowed afterward but global sea level continued to rise another 65 m at a more modest pace, reaching a level close to modern shorelines around 5000 BC (Rohde 2005).

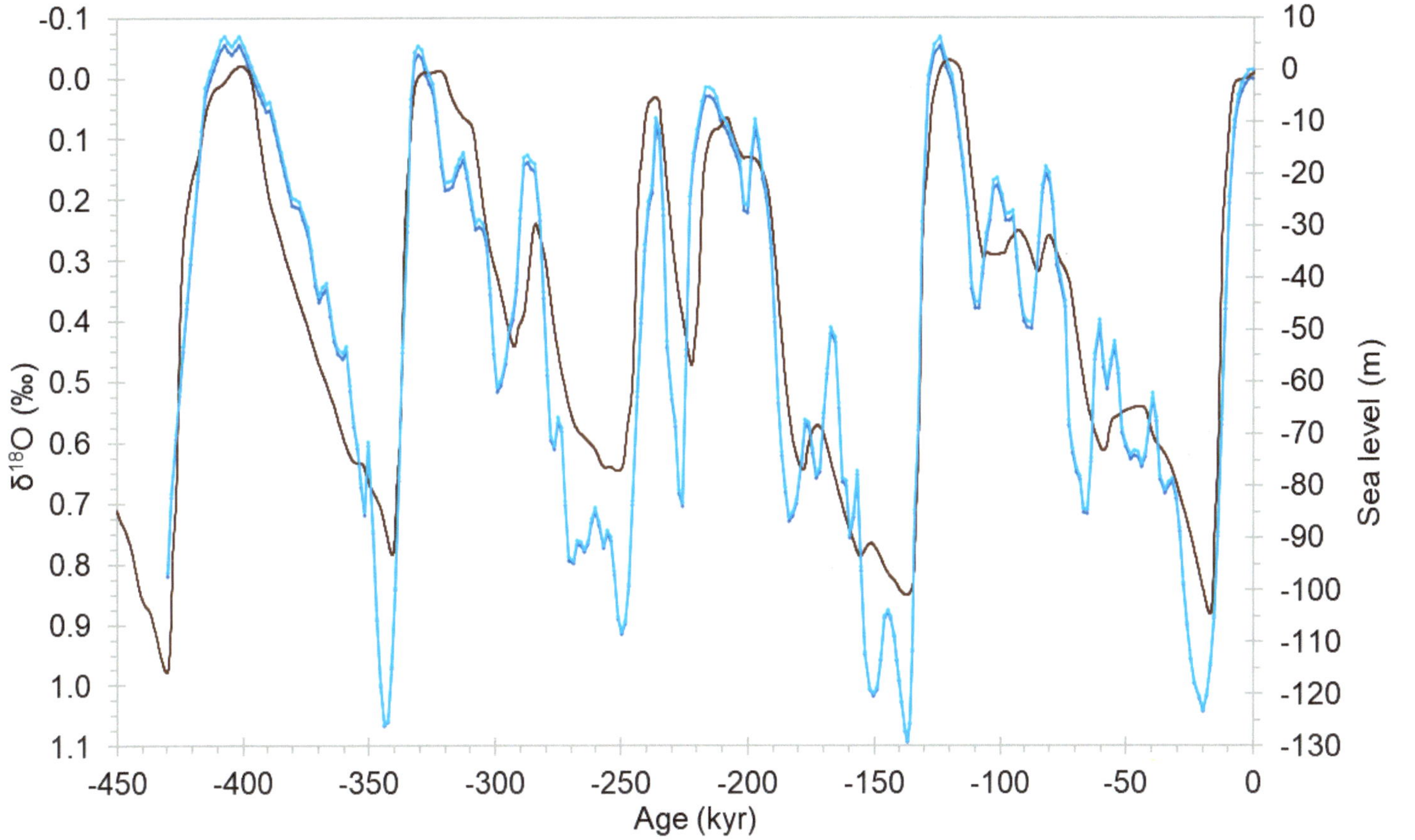

Figure 8D: δ¹⁸O and sea level over the last 450 kyr (DeBoer et al 2013, Waelbroeck et al 2002). The two blue curves show δ¹⁸O and sea level from ocean sediment records. The brown curve shows sea level inferred from 3-dimensional ice sheet volume simulations.

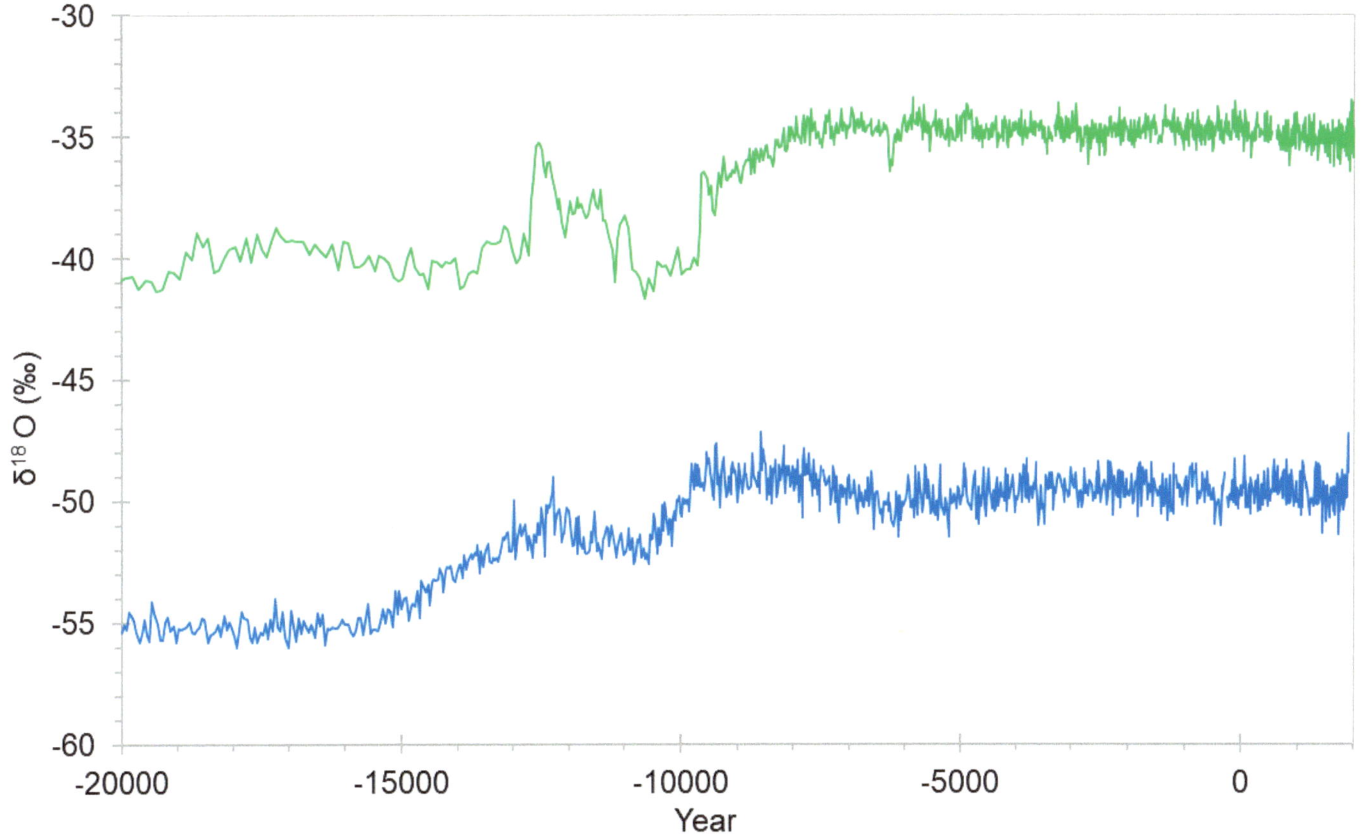

Figure 8E: δ¹⁸O data of the last 20 kyr from the Greenland Ice Sheet Project, in green, and the EPICA Dome C ice core in Antarctica, in blue (Grootes et al 1993, Meese et al 1994, Jouzel et al 2007).

Earth's Magnetic Field

This article is not about the migration of Earth's magnetic poles, which have moved a lot in the last four centuries (Newitt et al 2009). Yet, it's worth discussing this topic here so that the process of magnetic polar migration is not confused with the migration of Earth's rotational poles. In the early 17th century, the North Magnetic Pole (NMP) was likely somewhere north of Melville Island. By 1903, it had moved as far south as the Boothia Peninsula, where its location was first pinpointed by explorer Roald Amundsen. In 1947, Jack Clark and Paul Serson found that the NMP had moved ~300 km to a new location on Prince of Wales Island (Mandea 2022). Since then, the NMP has migrated significantly at an accelerating rate. In 1994, the NMP was at 73°15'N, 99°45'W, near the Isachsen weather station on Ellef Ringnes Island (Murray 2022). As of 2020, it was in the middle of the Arctic Ocean at 86°30'N 162°54'E, having moved ~1700 km toward Russia in 25 years (World Datacenter for Geomagnetism, Shi and Moldwin 2022).

The maximum separation of Earth's rotational and magnetic poles over the last few centuries has been ~20° of latitude, which is similar in scale to the magnetic declination of Jupiter. In addition to historical tracking of the NMP, paleogeographical changes in the position of the magnetic poles have been measured since 1954 using the geological record. (Creer et al 1954). The migration of the NMP to different locations on Earth's surface is now thought to be related to changes in the orientation of convection currents in the liquid outer core, which is what generates Earth's strong magnetic field lines (Weiss 2002).

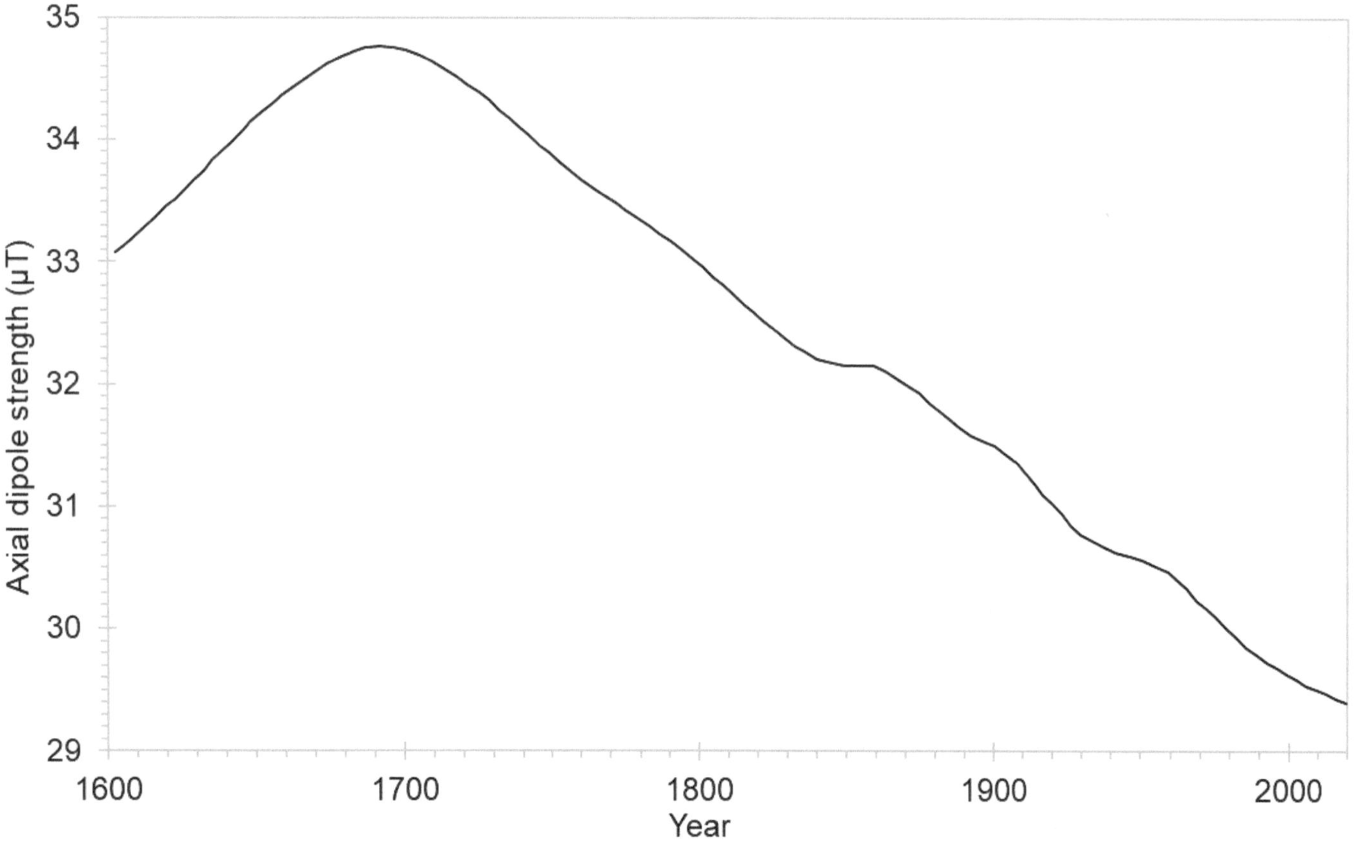

Figure 8F: Earth's magnetic field strength since 1600 (Jackson et al 2000, Korte and Constable 2011, Thebault et al 2015).

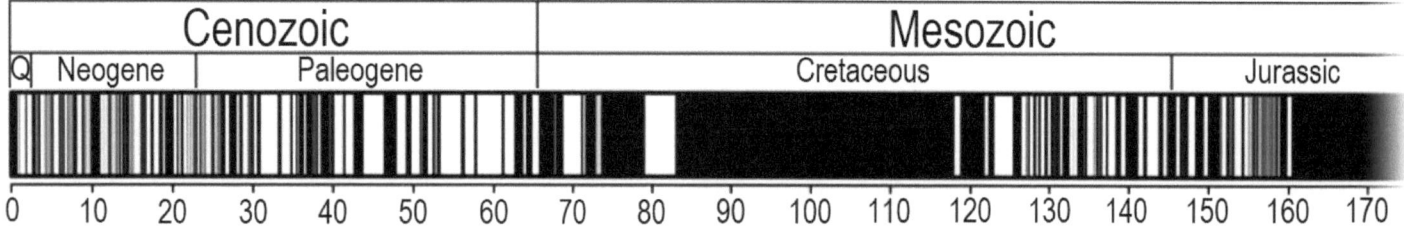

Figure 8G: Geomagnetic polarity over the last 170 Myr (Kent and Gradstein 1985, Cande and Kent 1995).

This article is also not about geomagnetic reversals. As you can see in figure 8F, Earth's magnetic field strength has declined about 15% since 1700 (Korte and Constable 2011). At various other times in the past, Earth's magnetic field strength has reached zero and its magnetic polarity has flipped, with the North Magnetic Pole becoming the South Magnetic Pole and vice versa. Earth's geomagnetic polarity record has been discerned from alternating magnetic stripes emerging out of both sides of divergent plate boundaries such as the Mid-Atlantic Ridge (Cox and Doell 1960, Vine and Matthews 1963, Morley and Larochelle 1964). The detailed magnetochronology in figure 8G shows that Earth's magnetic field was relatively stable in the Cretaceous period, with no magnetic reversals occurring between ~118 Myr and ~83 Myr ago. It has alternated more during the Cenozoic, and with an increased frequency in the last ~30 Myr (Opdyke and Channell 1996). During the Quaternary, geomagnetic reversals have occurred about once every 0.5 Myr, with the most recent one, the Brunhes-Matuyama event, happening ~781 kyr ago (Cande and Kent 1995, Gradstein et al 2004). Geomagnetic reversals take up to several thousand years to complete.

Geomagnetic reversals are probably not cataclysmic as they have not been linked to any major extinction events (Glassmeier and Vogt 2010). Earth's magnetic field is only ~1% as strong as a refrigerator magnet, so it's not strong enough to induce macroscopic effects on Earth's crust or interior. That being said, Earth's magnetic field is very effective at deflecting high-energy charged particles from the Sun or other stars, which prevents them from doing damage to life forms. A reduced magnetic field strength lets more external charged particles reach Earth's surface, but there isn't any clear evidence in the fossil record that this has been a problem for life on Earth. The Sun goes through regular magnetic polarity reversals about once every 11 years, which are also not associated with any major eruptive events or periods of instability. This concludes a brief overview of geomagnetodynamics, which is unrelated to migrations of Earth's rotational axis.

Climate Induced Polar Wander

True polar wander was discovered by Seth Chandler in the late 19th century (Newcomb 1891). It is a small nutation in Earth's axis that was predicted by Isaac Newton and Leonhard Euler based on their analyses of the dynamics of rotating rigid bodies (Euler 1765, Guicciardini 2005). True polar wander is disjoint from the obliquity and precessional cycles. In true polar wander, the location of the rotational poles changes compared to Earth's surface while Earth as a whole keeps spinning on the same sidereal axis. Although true polar wander has been discussed in scientific literature for decades, it has not been formally linked to any major events in geological history (Mueller 1969, Barnes et al 1983).

Imagine Earth as a rocky sphere freely rotating in space on an axis. Like other rotating bodies, it is driven to spin along an axis that fits its 3-dimensional distribution of mass optimally. When its distribution of mass changes, Earth as a whole is torqued to move so that it rotates along the most stable axis, which always runs through the center of mass. A freely-spinning figure skater who extends and holds one leg out to the side will be forced to rotate along a new axis that runs through the new center of mass somewhere between the extended leg and their torso.

It is known that large earthquakes and major construction projects have caused measurable effects on the speed of Earth's rotation and on the position of the geographic poles. The construction of the Three Gorges Dam from 1994 to 2010 enabled ~40 Gt of water from the Yangtze River to accumulate in a reservoir ~180 m above sea level which would have otherwise drained into the ocean (Wang et al 2002). This caused a measurable migration of the North Pole of ~2 cm and an increase in the length of the day by ~60 ns. As an amount of mass was moved further from Earth's center, the rate of rotation slowed in the same way a figure skater slows their rate of spin by stretching out their arms. In contrast, the large underwater earthquake that caused the 2004 Indian Ocean tsunami sped up Earth's rotation, subtracting ~2.7 μs from the length of the day, and moving the North Pole's position by ~2.5 cm (Chao and Gross 2005). This earthquake involved a subduction of the Indian Ocean Plate below the overriding Burma plate, which caused a slight compaction of the planet. This situation is analogous to a figure skater pulling

59

in their arms to increase their rate of spin. These two events both led to changes in the axis of rotation because the Earth needed to adjust its rotation to fit a different 3-dimesional distribution of mass.

More gradual and continuous wandering of Earth's rotational poles has been measured by scientists in real time. According to Surendra Ardhikari, the North Pole has drifted ~20 m over the last century (Adhikari and Ivins 2016). It was moving ~20 cm/year towards Toronto between 1900 and 2000, but then changed course and is now drifting more gently toward England. This measured motion of the North Pole has been attributed to gradual movements in Earth's core and mantle, continental drift, the melting of ice caps and post-glacial rebound (Lambeck 2005).

As continental ice sheets grow at high latitudes, they redistribute mass towards the rotational poles and away from the equatorial bulge. The ice sheets also have the effect of depressing the crust beneath them, bringing dense rocky material downward. When a large ice sheet melts and the pressure on the crust is relieved, the crust rebounds. Post-glacial rebound brings rocky material upwards in high-latitude regions to restore the surface to a more evenly distributed ellipsoidal layer. This redistribution of mass towards high latitude regions acts to counterbalance polar deglaciation's effect on Earth's moment of inertia. The part of Earth's surface with the largest measured post-glacial rebound is Hudson Bay (Paulson et al 2007). Ports set up on the shoreline of Hudson Bay just four centuries ago are now ~5 m above sea level and several kilometers inland because the crust in those areas is still rising more than five millennia after deglaciation (Pilon 1982, Lavoie et al 2012). During the Laurentide Ice Sheet's main phase of deglaciation, post-glacial rebound in the Hudson Bay region would have been as much as ~10 cm/year (Rasmussen et al 2006).

The proposed mechanism for climate induced polar migration at the end of the Pleistocene can be summarized as follows. An optimal alignment of the eccentricity and precessional cycles led to a maximal amount of summer ablation for Northern hemisphere ice sheets, which happens about once every 100 kyr. Albedo and ocean circulation based positive feedback mechanisms enhanced this melting, leading to rapid deglaciation and a rapid rise in sea level. The resulting change in Earth's center of mass precipitated a shift in its rotational axis. This shift in the rotational axis created additional positive

feedback that led to further melting, which led to further migration of the pole. The North Pole eventually settled into a new equilibrium position where new sheets of ice could accumulate. With this scenario in mind, a whole series of previous pole positions can be discerned from Pleistocene glacial records.

Central Greenland Location
(70°N, 50°W, 130 – 14.7 kyr ago)

With large continental ice sheets covering most of North America and Europe but lacking in Eastern Siberia, the most logical place for the position of the North Pole during the LGM would be close to 70°N, 50°W. This is ~2200 km south of the North Pole's current location. With this configuration, the Laurentide Ice Sheet would have swept over New England at ~58°N, the Great Lakes at ~56°N, and the Rocky Mountains at ~53°N. The Scandinavian ice sheet would have spread to the Northern British Isles at ~61°N, Northern Poland at ~57°N and Russia to the northeast of Moscow at ~52°N. Sea ice would have packed the North Atlantic and Arctic Oceans, with icebergs frequenting the coasts of Canada, Greenland, Iceland and Scandinavia. The LGM Cordilleran ice sheet extended from Southern Alaska to the northern part of Washington State, which would have represented the furthest reaches of Northern Hemisphere glaciation at ~49°N. Novaya Zemilya, a mountainous island separating the Barents and Kara seas was fully glaciated during the LGM but mainland Russia to the East of it lacked the same level of glaciation. Although high-altitude glaciers existed in the Alaska Range, the Brooke Range, the Verkhoyansk Range, the Taimyr Peninsula and the Kamchatka Peninsula, lower elevations of Eastern Siberia and Beringia were free of permanent ice.

This Central Greenland North Pole arrangement would place the LGM's South Pole off the coast of Antarctica towards Australia at 70°S, 130°E. No continental ice sheets existed in Australia during the LGM but extensive ice existed in the Tasmanian highlands and the western part of New Zealand's South Island at what would have been ~56°S. All of Antarctica was glaciated and much of Chile was covered by the Patagonian ice sheet, up to what would have been ~23°S latitude. That being said, Chilean coastal forests would have started near

~27°S and the Argentinian plains would have extended to Northern Tierra del Fuego at ~34°S (Veit and Garleff 1996, Villagran et al 2005). High-altitude glaciers covered most Andean plateaus during the LGM, even at the equator. For comparison, the Tibetan plateau currently holds glaciers at ~33°N. With all this in mind, a hypothetical scenario involving an LGM position of the North Pole in Central Greenland is believable.

The North Pole could have begun migrating from its LGM location in Central Greenland to its current position during the Bolling-Allerod warming event ~14.7 kyr ago. The Bolling-Allerod event was followed by meltwater pulse 1A, a sea level rise of ~20 m between ~14.7 kyr and ~13.5 kyr ago (Cronin 2012). As the Northern Hemisphere contains more of the world's continental landmass, this redistribution of water from high-latitude regions to the world's oceans would have caused Earth's center of mass to move towards the Southern Hemisphere. It would have also caused the Earth's rotation to slow down. It is not certain how long it would have taken the North Pole to reach a new equilibrium position, but the majority of any polar migration that may have occurred would have likely been complete by ~7 kyr ago, when global sea level was approaching modern day values. After meltwater pulse 1A came the Younger Dryas Cold Epoch, which lasted between ~13.5 kyr and ~11.7 kyr ago (Alley et al 1993). The Younger Dryas likely represented a response to vast amounts of icy cold meltwater from the Northern Hemisphere ice sheets being injected into the oceans (Alley et al 1993). The Younger Dryas ended ~11.7 kyr ago with a second intense warming event and sea level rise which initiated the Holocene Epoch.

About 100 kyr before the LGM was the Eemian interglacial, a warm period between ~130 kyr and ~115 kyr ago (Dahl-Jensen et al 2013). Many lines of evidence from the Northern Hemisphere suggest that the height of the Eemian was ~1 to 2°C warmer than today (Robinson et al 2011). Figure 8H shows a rough approximation of what the world was like during this time. Forests existed all the way up to Baffin Island and the northernmost part of Norway. Oak trees were growing in Finland and hippos were bathing in the river Thames (Van Kolfschoten 2000). Higher sea levels during the Eemian caused much of Greenland, Scandinavia, the West Siberian Plain and Florida to be underwater (Sakellariou 2015). East Antarctica was glaciated but most of West Antarctica, the part closer to South America,

was unglaciated, submerged and slowly rising due to post-glacial rebound. The paleogeographic layout during the Eemian provides further support to the notion that the North Pole could have been located in Central Greenland in the last segment of the Pleistocene.

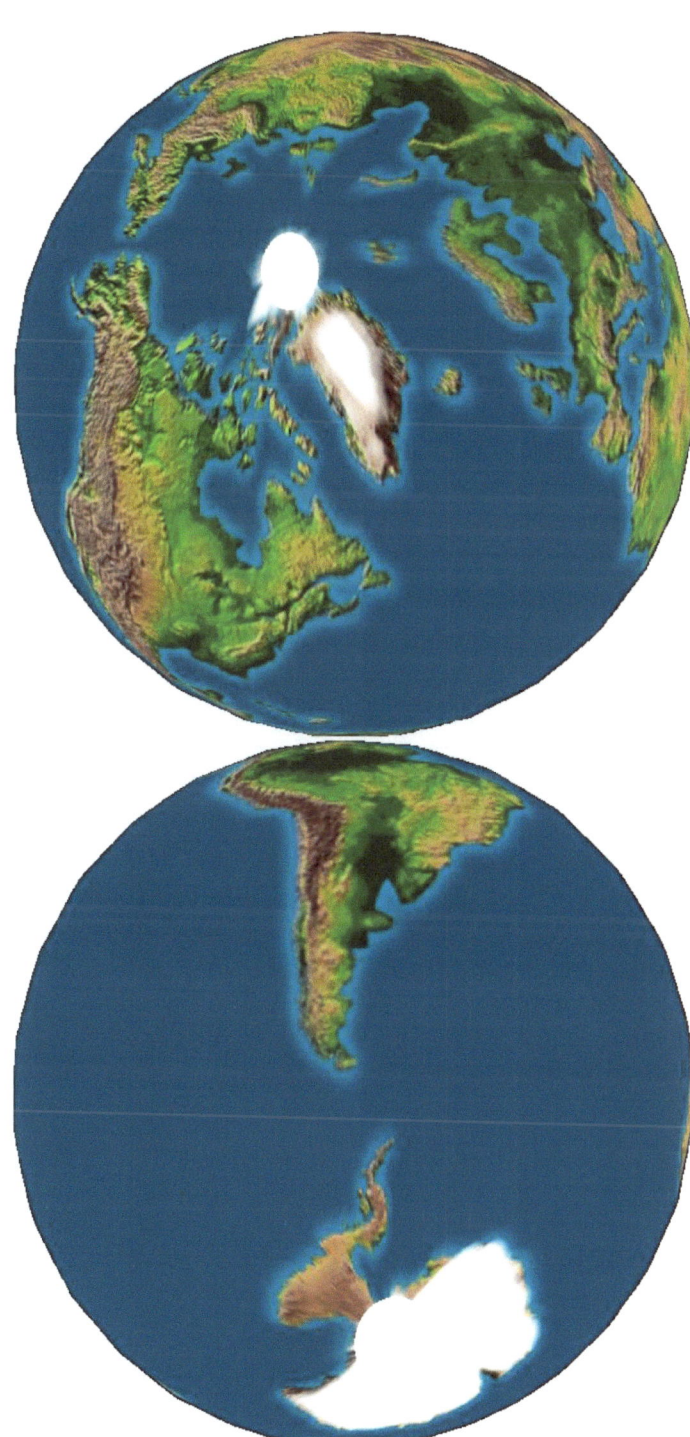

Figure 8H: Map of the world during the Eemian interglacial, ~120 kyr ago.

61

Svalbard Location
(77°N, 11°E, 243 – 130 kyr ago)

The Eemian interglacial was preceded by a lengthy glacial interval, which lasted from ~190 kyr to ~130 kyr ago (Landais et al 2003). The nadir of this glacial phase, centered ~140 kyr ago, is often dubbed the Penultimate Glacial Maximum (PGM) (Severinghaus et al 2006). The PGM coincided with the Moscow phase of the Saalian glaciation, when ice covered Scandinavia, the British Isles, and most of Poland and Belarus. The rest of Europe was mostly devoid of trees (Van Andel and Tzedakis 1996). The Saale ice sheet terminated near Moscow and covered the northern half of the Ural Mountains, the West Siberian Plain and the Taimyr Peninsula. The mountains of Kamchatka were glaciated and sea ice frequented the Sea of Okhotsk. Ice rafted debris could also be found in abundance on the Siberian side of the Arctic Ocean but was less common on the coasts of Eastern Canada than it was during the LGM (Manabe and Broccoli 1985). The Sahara Desert was enlarged, the rainforests of Equatorial Africa were greatly reduced and oak forests of the Mediterranean coast were wholly absent (Van Andel and Tzedakis 1996). Some sources claim that Europe and North America were stormier during the PGM and that North American ice sheets were much less extensive than their European counterparts (Colleoni et al 2016).

As the advance of the Laurentide Ice Sheet during the LGM would have erased evidence for previous glacial advances, it is hard to say exactly where North American ice sheets extended to during the PGM. Optically stimulated luminescence (OSL) dating techniques applied to fluvial deposits in the Glasford Formation of Illinois suggest that a pre-Wisconsin ice sheet covered much of North America between ~190 kyr and ~130 kyr ago (McKay and Berg 2008). However, the only volcanically-derived absolute dates that constrain pre-Wisconsin North American ice sheets are from the ~602-kyr-old Lava Creek B site (Hallberg 1986, Lisiecki and Raymo 2005). Dates obtained in earlier studies placed the most recent pre-Wisconsin glaciation, the Illionian, as far back as ~450 kyr ago, long before the PGM (Ericson and Wollin 1968, Hintze 1988, Wornardt and Vail 1991). Glacially buried sediments in British Columbia and Washington State also show evidence for a series of sizable pre-Cordilleran ice sheets that could have existed sometime before the LGM (Booth et al 2003).

Figure 8D's paleoclimatic record shows that sea level during the PGM was comparable to that of the LGM, suggesting that North America probably didn't have extensive continental glaciation during the PGM. This notion is reinforced by the fact that the coasts of Eastern Canada saw less ice rafted debris in the PGM than they did during the LGM. Since the European Saale Ice Sheet during the PGM was more extensive than its LGM counterpart and the temperature of Antarctica during the PGM was similar to that of the LGM, it would not have been possible to have Laurentide-sized ice sheets in North America during the PGM unless the sea level was significantly lower (Petit et al 1999). Due to this incompatibility, it's appropriate to say that older interpretations of the North American glacial record by D.B.Ericson, L.F.Hintze and W.W. Wornardt are deserving of further investigation and the more recently-derived OSL dates of the Glasford Formation deserve additional scrutiny.

Based on the arrangement of ice sheets during the PGM, the most sensible location for the North Pole's position between ~243 kyr and ~130 kyr ago is just southwest of the island of Svalbard, at approximately 77°N, 11°E. The relocation of the North Pole from this spot to its LGM position in Central Greenland could have been initiated at the beginning of the Eemian interglacial, ~127 kyr ago, when Earth's eccentricity was near a maximum and the Northern Hemisphere's summer solstice was near perihelion. A Svalbard location for the PGM North Pole can explain why the Labrador coast was ice-free at what would have been ~58°N, while Britain would have been mostly ice covered at ~64°N. Both Moscow and the Taimyr Peninsula would have both been at the edge of the ice sheet at ~67°N. One inconsistency, however, is that the Sea of Okhotsk would have contained abundant ice rafted debris at ~49°N. The PGM's South Pole would have been closer to New Zealand than it was during the LGM, while being about the same distance from Patagonia. This is consistent with the fact that a decent-sized ice sheet existed in the Southern Andes during its Santa Maria glacial phase, ~150 kyr ago (Porter 1981).

About 100 kyr before the PGM was the Aveley interglacial, which occurred between ~243 kyr and ~190 kyr ago on the British Isles (Lisiecki and Raymo 2005). The Aveley was roughly contemporaneous with the Domnitz and Gorka glacial retreats in Europe and Russia during marine isotope stage 7. During this time, Northern Europe and the

British Isles were largely unglaciated and populated with oak, hazel, hornbeam, water fern and Neanderthal societies that used Levallois stone tools (Schreve 2017). Figure 8D's paleotemperature record suggests that the Aveley was not as warm as the Eemian and Holocene interglacial phases, which could have been a result of destructive interference between the eccentricity and obliquity cycles. The warming phase that initiated the Aveley interglacial ~243 kyr ago may have spurred a migration of the North Pole from a previous position to its PGM location near Svalbard.

Earlier Pole Positions
(> 243 kyr ago)

Preceding the Aveley interglacial was the third latest glacial maximum (TGM), which took place ~250 kyr ago. According to figure 8D, sea level during that time was significantly higher than during the LGM and PGM. The TGM coincided with the early Wolstonian glaciation in Britain, the Funhe glaciation in Germany and the Vologda glaciation in Russia (Lisiecki and Raymo 2005). There were probably no extensive continental ice sheets in North America during the TGM as they could not exist alongside European ice sheets without a comparable drop in sea level.

Ice cover in the Northern Hemisphere during the TGM may have been limited to Greenland, the Canadian Arctic Archipelago, Iceland, Northern Britain, Northern Europe and Russia. The maximum advance of the Russian ice sheet during this glacial phase occurred during marine isotope stage 8, between ~300 kyr and ~243 kyr ago (Hughes et al 2020). There is no evidence for a synchronous glacial advance in Europe. A small European ice sheet coinciding with a large Russian ice sheet suggests a TGM North Pole location in a remote part of the Arctic Ocean around 85°N, 5°E. This would place the TGM South Pole close to the Queen Alexandra Range at 85°S, 175°W. Antarctic ice coverage during the TGM wouldn't have differed much in comparison to the LGM and PGM. The transition to this third latest pole position could have been initiated by a rapid warming phase ~337 kyr ago, in the early parts of Britain's Purfleet interglacial and Europe's Holstein interglacial (Lisiecki and Raymo 2005).

For the fourth latest pole position, between ~424 kyr and ~337 kyr ago, a location in the Davis Strait at approximately 75°N, 70°W makes the most sense. This was during the less extensive of two distinct episodes of the Wolstonian glaciation (Lee et al 2011). From the LGM pole position, this location would have been roughly the distance from the center of Wisconsin to the center of Illinois. A North Pole position in the Davis Strait would allow large continental ice sheets to exist in North America, including a precursor to the LGM's Cordilleran ice sheet, while not contributing to larger ice sheets in Europe. This general explanation for the fourth latest pole position is consistent with the sea level curve in figure 8D. The migration of the North Pole from a previous location to this Davis Strait location could have been initiated by a rapid warming phase at the beginning of the Hoxnian interglacial ~424 kyr ago (Lisiecki and Raymo 2005).

Preceding the Hoxnian interglacial was Britain's Anglian glaciation and Germany's Elster glaciation, which both began ~478 kyr ago (Lisiecki and Raymo 2005, Lee et al 2011, Bose et al 2012). The most sensible location for the position of the North Pole during this time would be near 75°N, 75°W. That would be just ~550 km west of the fourth latest pole position, which is about the distance from Illinois to Kansas. This configuration would allow large continental ice sheets to develop in both North America and Europe ~478 kyr ago. Figure 8D only goes back 450 kyr and it becomes much more difficult to estimate previous locations of the North Pole before this without a sea level curve.

The general pattern for the migration of the North Pole to a new location is that it tends to move directly away from the large ice sheets that are melting. This happens to be the direction in which Earth's center of mass is shifting. The movement of the pole from a Davis Strait location to a location in the Arctic Ocean ~337 kyr ago would have happened as continental ice sheets in North America melted alongside smaller Wolstonian ice sheets in Europe. When large ice sheets melted in Europe and Russia ~243 kyr ago, the pole would have moved from an Arctic Ocean location to somewhere near Svalbard. At the beginning of the Eemian interglacial ~130 kyr ago, the Saale ice sheets in Europe and Russia melted, and the pole would have moved from Svalbard to Greenland. Then, when the current interglacial started ~14.7 kyr ago, the Laurentide, Cordilleran and Weichselian ice sheets retreated and sent Earth's center of mass toward the unglaciated region of Beringia until it came to rest at its current location in the middle of the Arctic Ocean.

Calculation

The melting of the Laurentide, Cordilleran and Fennoscandian ice sheets at the end of the Pleistocene redistributed huge volumes of water from high northern latitudes into the world's oceans, raising the sea level by ~90 m and significantly altering Earth's 3-dimensional distribution of mass. To model this in a relatively simple way, we will imagine these three ice sheets together as the shell of a spherical cap centered on Hudson Bay at 60°N, 90°W and stretching about 20° of latitude in all directions. This arrangement covers an area of about 15 million km^2, which is similar in size to estimates for the combined area of the ice sheets during the LGM (Dyke et al 1982). Squeezing a 90 m difference in global sea level, about 32.5 quadrillion metric tons of water, into this hypothetical ice cap gives it an average thickness of ~2.2 km. This is consistent with estimates for the thickness of the Laurentide Ice Sheet during the LGM, which range between 1.2 and 4 km (Simon et al 2016).

Although Earth is approximately spherical in shape, the centrifugal force induced by its rotation causes the equator to bulge. This makes its equatorial radius ~0.33% larger than its polar radius, which is about double the elevation difference from the summit of Mount Everest to the depths of the Mariana Trench (King-Hele 1964, Fowles and Cassiday 2000). In that sense, Earth can be described as a slightly oblate spheroid. However, because the Northern Hemisphere contains a great deal more of the continental landmass and less ocean area than the Southern Hemisphere, Earth should actually be described as having an upside-down pear shape. Figure 8l attempts to illustrate this with greatly exaggerated features. The calyx of the pear represents the Arctic Ocean and the base of its stem represents the middle of Antarctica. When the ice sheets melted rapidly around the Pleistocene-Holocene transition, Earth's center of mass would have moved directly away from the Northern Hemisphere's ice sheets and towards Earth's center. The redistribution of the meltwater into the oceans would have then moved the center of mass southward, as more of the water would end up in the Southern Hemisphere. These movements of Earth's center of mass are represented in figure 8l by red arrows of greatly exaggerated magnitude. The present-day center of mass and Earth's center of mass during the LGM are represented by two X's.

Although the model and calculations shown here may seem like a gross simplification, they aim to provide a ballpark estimate for how much Earth's center of mass may have changed between the LGM and the present day. With this we can estimate the amount of polar migration that may have occurred around the Pleistocene-Holocene transition. This will be an overestimate which can constrain the maximum order of magnitude of such changes. First, we will calculate the difference between the current center of mass and the center of mass of a hypothetical Earth that has a 90 m sea level drop and $m_i = 32.5 \times 10^{18}$ kg of ice centered on 60°N, 90°W.

$$\Delta r_{CM} = \frac{(M_\oplus - m_i)(0) + m_i r_\oplus}{(M_\oplus - m_i) + m_i} \approx 34.6 \text{ m}$$

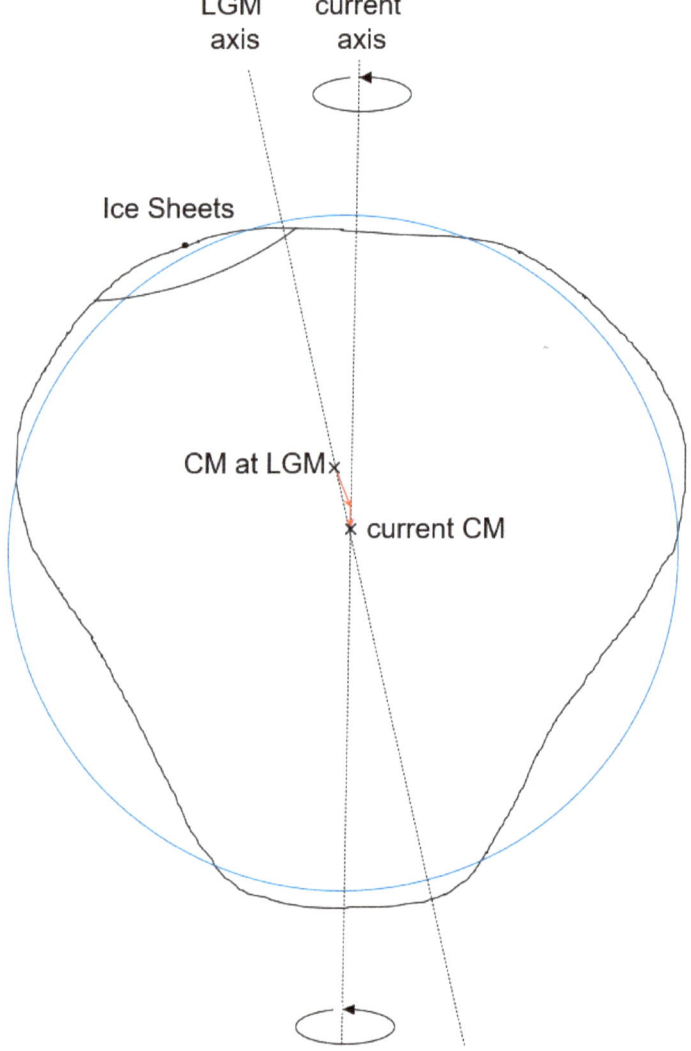

Figure 8l: An illustration of the change in Earth's center of mass from the LGM to today. This figure is greatly exaggerated.

As you can see from the calculation, the ice cap would move Earth's center of mass ~35 meters away from the ice sheets. Redistributing all of this mass into a bowl-shape that covers the Southern hemisphere moves the center of mass an additional ~17.3 m.

$$\Delta r_{CM} = \frac{(M_\oplus - m_i)(0) + \frac{1}{2}m_i r_\oplus}{(M_\oplus - m_i) + m_i} \approx 17.3 \text{ m}$$

Using trigonometry, we find that the total movement of Earth's center of mass would be ~50 m. In comparison, the construction of the Three Gorges Dam subtracted $m_w = 40 \times 10^{12}$ kg of water from the oceans and brought it ~180 m further away from the Earth's center (Wang et al 2002).

$$\Delta r_{CM} = \frac{(M_\oplus - m_i)(0) + m_w(r_\oplus + 180)}{(M_\oplus - m_i) + m_w} \approx 43 \text{ μm}$$

This would have moved Earth's center of mass by ~43 μm, which makes the melting of the hypothetical ice cap ~1.16 million times greater in magnitude. Since the Three Gorges Dam produced a measurable change in the Earth's axis of ~2 cm, linear extrapolation produces a ballpark estimate of ~23 km for the polar shift at the Pleistocene-Holocene transition. We must be reminded that this is an overestimate. It is only about a hundredth of the 2200 km and 20 degrees latitude required to bring the North Pole from Central Greenland to its current position.

Discussion

Since the calculation yielded an estimate of 23 km for polar migration around the Pleistocene-Holocene transition, it is understandable for one to express doubt for large movements of the North Pole occurring during the Pleistocene. That being said, there are two possibilities to consider. Either there has been no major polar movements and the scenario described here is nonsense, or the polar shift scenario is realistic and something is missing in the theory or the calculations. Accepting the first option leaves the ice-free Beringia problem unresolved. The latter option can explain the asymmetrical distribution of ice sheets during the LGM and does not contradict any theoretical aspects of Milankovitch

cycles. Its only major drawback is that the calculated movement of the pole is about two orders of magnitude smaller than the expected value.

The polar migration hypothesis provides an additional positive feedback mechanism to explain the exceptionally rapid increase in temperature and sea level between ~14.7 kyr and ~11.7 kyr ago. A change in the latitude of continental ice sheets brings them to regions with more annual sunlight, which melts more ice, which induces further polar migration. However, this still can't explain why the calculated value of the hypothetical polar migration is 23 km or less.

Changes in Earth's center of mass during the late Pleistocene and early Holocene were essentially controlled by three main processes: deglaciation, sea level rise and post-glacial rebound. Because post-glacial rebound returns mass to regions that have deglaciated, it counterbalances the effect of deglaciation on Earth's center of mass. The balance between deglaciation and post-glacial rebound may explain how a major shift in the poles could occur during the warming phases at the end of glacial periods but not intermittently throughout the Holocene.

The idea of polar migration during the Pleistocene may even help explain some as-yet-unresolved aspects of Milankovitch theory and the Quaternary ice ages in general. It isn't at odds with the notion that transitions from glacial to interglacial phases happen abruptly yet ice accumulates more gradually during lengthy glacial phases. Polar migration during the Pleistocene provides additional context for discussing why the ~100 kyr cycle became pronounced in the paleotemperature record ~1.0 Myr ago. The poleshifts described here would have been induced by the rapid melting of continental ice sheets and sea ice after the mid-Pleistocene transition. Therefore, about eight of these poleshifts could have taken place, all separated by ~100 kyr on average. With all these potential relocations, it appears that the North Pole has not ventured further than about 30° latitude from the center of Greenland and that the South Pole hasn't ventured far from Antarctica. According to the analysis presented here, the North Pole's current location is closer to Alaska and Eastern Siberia than any previous pole positions. This may serve to explain the lack of evidence for previous extensive continental ice sheets in those regions.

Conclusions

This article presented a potential scenario for a shift in the position of the Earth's rotational poles as a result of deglaciation at the end of the Pleistocene. A migration of the North Pole at the end of the Pleistocene would explain why Canada was almost completely covered in ice during the LGM but Eastern Siberia and Alaska were not. The alternative explanation relies on dubious theoretical claims that ice sheets didn't cover Beringia because it lacked sufficient precipitation, which clearly contradicts paleontological and paleobotanical evidence.

A series of movements of the North Pole over the Pleistocene have been discerned, with previous pole positions in Central Greenland, Svalbard, the Arctic Ocean and the Davis Strait. Most of the movement from the North Pole's LGM position around 70°N, 50°W to its current location probably occurred around the Pleistocene-Holocene transition ~14.7 kyr to ~11.7 kyr ago. As the ice caps melted and Earth's center of mass shifted, it slipped into a new dynamical equilibrium with a new rotational axis.

The idea that the North Pole could have been abruptly relocated several times during the Pleistocene is compatible with standard Milankovitch theory and can explain some aspects of the Quaternary ice ages that have been previously overlooked. A relocation of the poles at the end of the Pleistocene also provides an additional positive feedback mechanism to explain the exceptionally rapid warming from ~14.7 kyr to ~11.7 kyr ago.

The calculations presented here are not able to prove that the melting of the LGM ice sheets could have moved the North Pole's position from Central Greenland to where it is today as the calculated value of 23 km for polar migration is only ~1% of this distance. Therefore, in order to conclude that a significant degree of polar migration took place at the end of the Pleistocene, modifications to the theoretical and/or calculational framework presented in this article would be necessary.

Acknowledgments

Paul McNeil is thanked for sharing his knowledge of geology and providing insights that helped generate this article.

References

Adhikari,S.,Ivins,E.R.(2016).Climate-driven polar motion: 2003–2015.*Science advances*,2(4),e1501693.

Agassiz,L.,Bettannier,J.(1840).*Etudes sur Les Glaciers:Atlas*. Nicolet.

Alley,R.B.,Meese,D.A.,Shuman,C.A.,Gow,A.J.,Taylor,K.C., Grootes,P.M.,et al.(1993).Abrupt increase in Greenland snow accumulation at the end of the Younger Dryas event.*Nature*,362(6420),527-529.

Barnes,R.T.H.,Hide,R.,White,A.A.,Wilson,C.A.(1983).Atmospheric angular momentum fluctuations,length-of-day changes and polar motion.*Proceedings of the Royal Society of London.A.Mathematical and Physical Sciences*,387 (1792),31-73.

Bartoli,G.,Sarnthein,M.,Weinelt,M.,Erlenkeuser,H.,Garbe-Schönberg,D.,Lea,D.W.(2005).Final closure of Panama and the onset of northern hemisphere glaciation.*Earth and Planetary Science Letters*,237(1-2),33-44.

Bird,M.I.,Taylor,D.,Hunt,C.(2005).Palaeoenvironments of insular Southeast Asia during the Last Glacial Period:a savanna corridor in Sundaland?.*Quaternary Science Reviews*,24(20-21),2228-2242.

Booth,D.B.,Troost,K.G.,Clague,J.J.,Waitt,R.B.(2003).The Cordilleran ice sheet.*Developments in Quaternary Sciences*,1,17-43.

Böse,M.,Lüthgens,C.,Lee,J.R.,Rose,J.(2012).Quaternary glaciations of northern Europe.*Quaternary Science Reviews*, 44,1-25.

Buis,A.(2020).Milankovitch (orbital) cycles and their role in earth's climate.*NASA Climate*,27.

Carlson,A.E.,Tarasov,L.,Pico,T.(2018).Rapid Laurentide ice-sheet advance towards southern last glacial maximum limit during marine isotope stage 3.*Quaternary Science Reviews*,196,118-123.

Cande,S.C.,Kent,D.V.(1995).Revised calibration of the geomagnetic polarity timescale for the Late Cretaceous and Cenozoic.*Journal of Geophysical Research:Solid Earth*, 100(B4),6093-6095.

Chao,B.F.,Gross,R.S.(2005).Did the 26 December 2004 Sumatra, Indonesia, earthquake disrupt the Earth's rotation as the mass media have said?*EOS,Transactions American Geophysical Union*.Vol.86(1).

Colleoni,F.,Wekerle,C.,Näslund,J.O.,Brandefelt,J.,Masina,S. (2016).Constraint on the penultimate glacial maximum Northern Hemisphere ice topography (≈140 kyrs BP).*Quaternary Science Reviews*,137,97-112.

Cox,A.,Doell,R.R.(1960).Review of paleomagnetism.*Geological Society of America Bulletin*,*71*(6),645-768.

Creer,K.M.,Irving,E.,Runcorn,S.K.(1954).The direction of the geomagnetic field in remote epochs in Great Britain.*Journal of geomagnetism and geoelectricity*,*6*(4),163-168.

Croll,J.(1885).*Climate and time in their geological relations:a theory of secular changes of the Earth's climate*.A.and C. Black.

Cronin,T.M.(2012).Rapid sea-level rise.*Quaternary Science Reviews*,*56*,11-30.

Dahl-Jensen,D.et al.(2013)"Eemian interglacial reconstructed from a Greenland folded ice core."*Nature* 493,no.7433: 489-494.

De Boer,B.,Vande Wal,R.S.W.,Lourens,L.J.,Bintanja,R., Reerink,T.J.(2013).A continuous simulation of global ice volume over the past 1 million years with 3-D ice-sheet models.*Climate Dynamics*,*41*,1365-1384.

Dyke,A.S.,Dredge,L.A.,Vincent,J.S.(1982).Configuration and dynamics of the Laurentide Ice Sheet during the Late Wisconsin maximum.*Géographie physique et Quaternaire*, *36*(1),5-14.

Ericson,D.B.,Wollin,G.(1968).Pleistocene Climates and Chronology in Deep-Sea Sediments:Magnetic reversals give a time scale of 2 million years for a complete Pleistocene with four glaciations.*Science*,*162*(3859),1227-1234.

Esmark,J.(1824).Bidrag til vor jordklodes historie.*Magazin for Naturvidenskaberne*,*2*(1),28-49.

Euler,L.(1765).Du mouvement de rotation des corps solides autour d'un axe variable.*Mémoires de l'académie des sciences de Berlin*,154-193.

Felzer,B.(2001).Climate impacts of an ice sheet in East Siberia during the Last Glacial Maximum.*Quaternary Science Reviews*,*20*(1-3),437-447.

Fowles,G.,Cassiday,G.(2000).Analytical Mechanics.*Brooks Cole*.

Geomagnetic and Magnetic Poles.*World Data Centre for Geomagnetism,Kyoto,*(website).

Glassmeier,K.H.,Vogt,J.(2010).Magnetic Polarity Transitions and Biospheric Effects.*Space science reviews*,*155*.

Gradstein,F.M.,Ogg,J.G.Smith,A.G.(Eds.).(2004).A geologic time scale 2004(Vol. 86).*Cambridge University Press*.

Grootes,P.M.,Stuiver,M.,White,J.W.C.,Johnsen,S.,Jouzel,J. (1993).Comparison of oxygen isotope records from the GISP2 and GRIP Greenland ice cores.*nature*,*366*(6455), 552-554.

Guicciardini,N.(2005).Isaac Newton,philosophiae naturalis principia mathematica,(1687).In *Landmark Writings in Western Mathematics 1640-1940(59-87)*.Elsevier Science.

Hallberg,G.R.(1986).Pre-Wisconsin glacial stratigraphy of the central plains region in Iowa, Nebraska, Kansas, and Missouri.*Quaternary Science Reviews*,*5*,11-15.

Hays,J.D.,Imbrie,J.,Shackleton,N.J.(1976).Variations in the Earth's Orbit:Pacemaker of the Ice Ages:For 500,000 years,major climatic changes have followed variations in obliquity and precession.*Science*,*194*(4270),1121-1132.

Hintze,L.F.(1988).Geologic history of Utah.*Brigham Young Univ.Geology Studies,Spec.Publ*,7,202p.

Hoffecker,J.F.,Elias,S.A.(2007).*Human ecology of Beringia.* Columbia University Press.

Hopkins,D.M.(1967).*The Bering land bridge* (Vol.3).Stanford University Press.

Hughes,P.D.,Gibbard,P.L.,Ehlers,J.(2020).The"missing.glaciations"of.the.Middle.Pleistocene.*Quaternary.Research*,*96*, 161-183.

Hultén,E.(1937).Outline of the history of arctic and boreal biota during the Quaternary period.

Jackson,A.,Jonkers,A.R.,Walker,M.R.(2000).Four centuries of geomagnetic secular variation from historical records. *Philosophical Transactions of the Royal Society of London. Series A:Mathematical, Physical and Engineering Sciences,* 358(1768),957-990.

Jouzel,J.,Masson-Delmotte,V.,Cattani,O.,Dreyfus,G.,Falourd, S.,Hoffmann,G.et al.(2007).EPICA Dome C ice core 800 kyr deuterium data and temperature estimates.*IGBP PAGES/World Data Center for Paleoclimatology data contribution series*,*91*.

Kent,D.V.,Gradstein,F.M.(1985).A Cretaceous and Jurassic geochronology.*Geological Society of America Bulletin*,*96* (11),1419-1427.

King-Hele,D.G.(1964).The shape of the earth.*The.Journal of Navigation*,*17*(1),1-16.

Korte,M.,Constable,C.(2011).Improving geomagnetic field reconstructions for 0–3 ka.*Physics of the Earth and Planetary Interiors*,*188*(3-4),247-259.

Lambeck,K.(2005).The Earth's variable rotation:geophysical causes and consequences.*Cambridge University Press*.

Landais,A.,Chappellaz,J.,Delmotte,M.,Jouzel,J.,Blunier,T., Bourg,C.,Steffensen,J.P.(2003).A tentative reconstruction of the last interglacial and glacial inception in Greenland based on new gas measurements in the Greenland Ice Core Project(GRIP)ice core.*Journal of Geophysical Research:Atmospheres*,*108*(D18).

Laskar,J.,Fienga,A.,Gastineau,M.,Manche,H.(2011).La2010: a new orbital solution for the long-term motion of the Earth.*Astronomy & Astrophysics*,*532*,A89.

Lavoie,C.,Allard,M.,Hill,P.R.(2002).Holocene deltaic sedimentation along an emerging coast:Nastapoka River delta,eastern Hudson Bay,Quebec.*Canadian Journal of Earth Sciences*,*39*(4),505-518.

Lavoie,C.,Allard,M.,Duhamel,D.(2012).Deglaciation landforms and C-14 chronology of the Lac Guillaume-Delisle area, eastern Hudson Bay:a report on field evidence.*Geomorphology*,*159*,142-155.

Lee,J.R.,Rose,J.,Hamblin,R.J.,Moorlock,B.S.,Riding,J.B.,Phillips,E.,Candy,I.(2011).The Glacial History of the British Isles during the Early and Middle Pleistocene:Implications for the long-term development of the British Ice Sheet. In *Developments in Quaternary Sciences*(Vol.15,59-74). Elsevier.

Lisiecki,L.E.,Raymo,M.E.(2005).A Pliocene-Pleistocene stack of 57 globally distributed benthic $\delta^{18}O$ records.*Paleoceanography,20*(1).

Liu,Z.,Pagani,M.,Zinniker,D.,DeConto,R.,Huber,M.,Brinkhuis,H.,Pearson,A.(2009).Global cooling during the Eocene-Oligocene climate transition.*Science,323*(5918),1187-1190.

Manabe,S.,Broccoli,A.J.(1985).The influence of continental ice sheets on the climate of an ice age.*Journal of Geophysical Research:Atmospheres,90*(D1),2167-2190.

Mandea,M.(2022).Geomagnetic and Magnetic Poles.In *The Magnetic Declination:A History of the Compass*(103-112). Cham:Springer International Publishing.

McKay III,E.D.,Berg,R.C.(2008).Optical ages spanning two glacial-interglacial cycles from deposits of the ancient Mississippi River,north-central Illinois.In *Abstracts with Programs-Geological Society of America,North-Central Section,42nd annual meeting*(78).

Meese,D.,Alley,R.,Gow,T.,Grootes,P.M.,Mayewski,P.,Ram, M.,Zielinski,G.(1994).Preliminary depth-age scale of the GISP2 ice core.CRREL Special Report 94-1.*US Army Cold Regions Research and Engineering Laboratory,Hanover,NH,66*.

Milankovitch,M.M.(1941).Canon of insolation and the iceage problem.*Koniglich Serbische Akademice Beograd Special Publication,132*.

Morley,L.W.,Larochelle,A.(1964).Paleomagnetism as a means of dating geological events.*Geochronology in Canada,8*,39-51.

Mueller,I.I.(1969).Spherical and practical astronomy,as applied to geodesy.*New York*.

Murray,D.(2022).The 46th Reconnaissance Squadron:Arctic Exploration and Questions of Sovereignty in the Early Cold War.*The Northern Mariner/Le marin du nord,32*(1), 39-70.

Newcomb,S.(1891).On the periodic variation of latitude, and the observations with the Washington primevertical transit. *The Astronomical Journal,11*,81-82.

Newitt,L.R.,Chulliat,A.,Orgeval,J.J.(2009).Location of the north magnetic pole in April 2007.*Earth, planets and space,61*,703-710.

Opdyke,M.D.,Channell,J.E.(1996).Magnetic stratigraphy.*Academic press*.

Paulson,A.,Zhong,S.,Wahr,J.(2007).Inference of mantle viscosity from GRACE and relative sea level data.*Geophysical Journal International,171*(2),497-508.

Petit,J.R.,Jouzel,J.,Raynaud,D.,Barkov,N.I.,Barnola,J.M.,Basile,I.,Stievenard,M.(1999).Climate and atmospheric history of the past 420000 years from the Vostok ice core, Antarctica.*Nature,399*(6735),429-436.

Pico,T.,Birch,L.,Weisenberg,J.,Mitrovica,J.X.(2018).Refining the Laurentide Ice Sheet at Marine Isotope Stage 3:A data-based approach combining glacial isostatic simulations with a dynamic ice model.*Quaternary Science Reviews,195*,171-179.

Pilon,J.L.(1982).Fort Severn Land Use and Occupancy Study,PART.

Pisias,N.G.,Moore Jr,T.C.(1981).The evolution of Pleistocene climate:a time series approach.*Earth and Planetary Science Letters,52*(2),450-458.

Poore,R.Z.,Williams Jr,R.S.,Tracey,C.(2000).Sea level and climate.*US Geological Survey Fact Sheet,2*(00),1-2.

Porter,S.C.(1981).Pleistocene glaciation in the southern Lake District of Chile.*Quaternary Research,16*(3),263-292.

Rasmussen,S.O.,Andersen,K.K.,Svensson,A.M.,Steffensen J.P.,Vinther,B.M.,Clausen,H.B.,Ruth,U.(2006).A new Greenland ice core chronology for the last glacial termination.*Journal of Geophysical Research:Atmospheres, 111*(D6).

Robinson,A.,Calov,R.,Ganopolski,A.(2011).Greenland ice sheet model parameters constrained using simulations of the Eemian Interglacial.*Climate of the Past,7*(2),381-396.

Rohde,R.A.(2005).Global warming art project.

Sakellariou,V.(2015).Vivien Gornitz:Rising seas:Past, present,future.*New York:Columbia University Press*.

Schreve,D.(2017).Neither hot nor cold but dry:A Northwest European view of Neanderthal environments in late MIS 7 and beyond.In *Crossing the Human Threshold*(192-214). Routledge.

Schulz,K.G.,Zeebe,R.E.(2006).Pleistocene glacial terminations triggered by synchronous changes in Southern and Northern Hemisphere insolation:The insolation canon hypothesis.*Earth and Planetary Science Letters,249*(3-4), 326-336.

Seidelmann,P.K.(Ed.).(1992).Explanatory supplement to the astronomical almanac.*University Science Books*.

Severinghaus,J.P.,Kawamura,K.,Headly,M.(2006,December).Evidence of deep air convection in firn at Vostok,Antarctica in the Penultimate Glacial Maximum from precise measurements of Kr isotopes.In *AGU Fall Meeting Abstracts*(Vol. 2006,U33C-03).

Shi,Y.,Moldwin,M.B.(2022).Interhemispheric Asymmetries in Magnetosphere and Ionosphere Magnetic Field Residuals Between Swarm Observations and Earth Magnetic Field Models.*Journal of Geophysical Research:Space Physics, 127*(3),e2021JA030190.

Simon,K.M.,James,T.S.,Henton,J.A.,Dyke,A.S.(2016).A glacial isostatic adjustment model for the central and northern Laurentide Ice Sheet based on relative sea level and GPS measurements.*Geophysical Journal International*,*205*(3), 1618-1636.

Thébault,E.,Finlay,C.C.,Beggan,C.D.,Alken,P.,Aubert,J.,Barrois,O.,Zvereva,T.(2015).International geomagnetic reference field:the 12th generation.*Earth,Planets and Space*,*67*,1-19.

Tiedemann,R.,Sarnthein,M.,Shackleton,N.J.(1994).Astronomic timescale for the Pliocene Atlantic $\delta^{18}O$ and dust flux records of Ocean Drilling Program Site 659.*Paleoceanography*,*9*(4),619-638.

Tierney,J.E.,Zhu,J.,King,J.,Malevich,S.B.,Hakim,G.J., Poulsen,C.J.(2020).Glacial cooling and climate sensitivity revisited.*Nature*,*584*(7822),569-573.

VanAndel,T.H.,Tzedakis,P.C.(1996).Palaeolithic landscapes of Europe and environs,150,000-25,000 years ago:an overview.*Quaternary science reviews*,*15*(5-6),481-500.

Van Kolfschoten,T.(2000).The Eemian mammal fauna of central Europe.*Netherlands Journal of Geosciences*,*79*(2-3),269-281.

Veit,H.,Garleff,K.(1996).Evolución del paisaje cuaternario y los suelos en Chile central-sur.*Ecología de los bosques nativos de Chile*,29-49.

Villagrán,C.,Hinojosa,L.F.,Llorente-Bousquets,J.,Morrone, J.J.(2005).Esquema biogeográfico de Chile.*Regionalización Biogeográfica en Iberoamérica y Tópicos Afines: Primeras Jornadas Biogeográficas de la Red Iberoamericana de Biogeografía y Entomología Sistemática. Las Prensas de Ciencias,UNAM,Mexico City*,551-557.

Vine,F.J.,Matthews,D.H.(1963).Magnetic anomalies over oceanic ridges.*Nature*,*199*(4897),947-49.

Voous,K.H.(1973).Proceedings of the 15th International Ornithological Congress,The Hague,The Netherlands 30 August–5 September 1970.*Brill Archive*.

Waelbroeck,C.,Labeyrie,L.,Michel,E.,Duplessy,J.C.,Mcmanus,J.F.,Lambeck,K.,Labracherie,M.(2002).Sea-level and deep water temperature changes derived from benthic foraminifera isotopic records.*Quaternary science reviews*,*21*(1-3),295-305.

Walker,M.,Johnsen,S.,Rasmussen,S.O.,Popp,T.,Steffensen,J.P.,Gibbard,P.,Schwander,J.(2009).Formal definition and dating of the GSSP (Global Stratotype Section and Point) for the base of the Holocene using the Greenland NGRIP ice core, and selected auxiliary records.*Journal of Quaternary Science:Published for the Quaternary Research Association*,*24*(1),3-17.

Wang,H.,Hsu,H.T.,Zhu,Y.Z.(2002).Prediction of surface horizontal displacements, and gravity and tilt changes caused by filling the Three Gorges Reservoir.*Journal of Geodesy*, *76*,105-114.

Ward,W.R.(1982).Comments on the long-term stability of the Earth's obliquity.*Icarus*,*50*(2-3),444-448.

Weiss,N.(2002).Dynamos in planets,stars and galaxies.*Astronomy & Geophysics*,*43*(3),3-9.

Wornardt.W.W.,Vail,P.R.(1991)Revision of the Plio-Pleistocene Cycles and their Application to Sequence Stratigraphy and Shelf and Slope Sediments in the Gulf of Mexico.*Gulf Coast Association of Geological Societies Transactions*.v.41,719-744.

Yu,J.,Menviel,L.,Jin,Z.D.,Thornalley,D.J.R.,Foster,G.L., Rohling,E.J.,Roberts,A.P.(2019).More efficient North Atlantic carbon pump during the last glacial maximum.*Nature communications*,*10*(1),2170.

Zazula,G.D.,Froese,D.G.,Schweger,C.E.,Mathewes,R.W., Beaudoin,A.B.,Telka,A.M.,Westgate,J.A.(2003).Ice-age steppe vegetation in east Beringia.*Nature*,*423*(6940),603.

Do planetary alignments affect solar cycles, climate and economic productivity?

Abstract

Solar activity and sunspot counts go through a range of quasi-periodic oscillations. The largest of these is based on an average period of ~11 years. A multitude of other solar periodicities contribute to regular variations in sunspot activity. These oscillations appear to be based on the repeating and beating tidal influence of planets on the Sun's surface or tachocline. In this article, several commonly reported solar periodicities are explored, and dates related to individual solar cycles are compared to notable planetary alignment configurations. Based on this analysis, local maxima in combined planetary tidal influence tend to correspond with periods of decreased sunspot activity, which in turn tend to relate to times of terrestrial climatic deterioration and downturns in economic prosperity. Here, a few possible connections between historically significant solar, climatic, economic and political events and individual solar periodicities are explored through a series of tables. This is an atypical approach, as most studies involving solar periodicities focus on spectral and wavelet analysis and not on the timing of planetary alignments. In this study, a mathematical model of total solar activity is developed from an overview of these prominent solar periodicities and important planetary alignment events. The model is composed of a series of repeating negative delta functions of various amplitudes, breadths, periods and phase shifts. When this model is compared to historical group sunspot number counts from 1705 to 1995, it yields a Pearson correlation coefficient of 93%. Through this novel approach, this study attempts to show that some recurrent patterns in history are guided by important planetary alignment patterns.

Introduction

Solar chronologies from Chinese astronomical records date all the way back to the 9th century BC (Zhen-tao 1980). The introduction of new telescopic techniques in the early 17th century expanded and enhanced the historical sunspot record. Around the year 1610, several European astronomers began writing about sunspots including the Fabricious brothers of Holland, Englishman Thomas Harriot, German Christoph Scheiner and Italian Galileo Galilei (Vokhmyanin 2020). The late 18th century English astronomer William Herschel described a regular 11-year periodicity in sunspot activity and claimed that it had an influence on the price of wheat (Herschel 1801). In the mid-19th century, Swiss astronomer Rudolf Wolf suggested that Herschel's 11-year variation in sunspots could depend on the combined gravitational influence of Venus, Earth, Jupiter and Saturn (Wolf 1859). Other astronomers such as John Henry Poynting, disputed such ideas, claiming that a lack of supporting evidence and a series of failed predictions made such connections speculative at best (Poynting

1884, Love 2013). However, a lack of alternative explanations for periodic oscillations in solar activity continues to warrant investigation of potential relationships between sunspot activity and planetary alignments.

This article explores potential connections between notable types of planetary alignments and some prominent periodicities in the sunspot record. At times this analysis will involve comparisons with historically significant climatic, economic and political events. It aims to be less mathematically exhaustive than parallel studies of planetary influences on solar variation, and concentrates more on comparing maxima and minima of commonly reported solar periodicities to important planetary alignments (Baidolda 2017). The characteristics of the ~11-year solar periodicity are explained, but this article focuses more on longer-term variations in sunspot activity with multi-decadal and multi-centurial timescales. This article makes the bold yet obvious assumption that if planetary alignments influence solar activity, then this in turn influences Earth's climate. It suggests that, in general, planetary tidal maxima tend to correspond with solar activity minima, which in turn coincide with times of agricultural and economic hardship among human societies.

Physical Mechanisms

The first inquiry into physical mechanisms for the ~11-year sunspot cycle and other solar periodicities was presented by Eugene Parker in 1958 (Parker 1958). Parker suggested that planetary tides periodically perturb the Sun's surface magnetic field and shear the Sun's tachocline, the boundary between the radiative and convective zones of the Sun's interior. He claimed that this in turn causes spurs in sunspot activity, as well as pulses in the corona and solar wind. In 1961, Horace Babcock offered a more sophisticated model that involved solar magnetic flux being transported from the surface to the tachocline by meridional circulation rather than by diffusion (Babcock 1961). In 1980, Edward Spiegel and Nigel Weiss proposed that the dynamo process occurs exclusively in the tachocline and the lower convective zone, and that observed surficial processes are merely a scrambled version of the dynamic processes occurring underneath (Spiegel and Weiss 1980). In 2010, Charles Wolff and Paul Patrone hypothesized that the tidal influence of the planets on the Sun could affect nuclear reactions in its core (Wolff and Patrone 2010). In 2012, Jose Abreu suggested that repeating tidal interactions between the planets and the Sun could cause a small migration of the tachocline (Abreu et al 2012). This shift in the tachocline could in turn lead to positive feedback loops and affect hydrogen fusion processes in the Sun's core. More recent papers have proposed that sunspot changes may result from nonlinear chaotic dynamo processes in the solar interior (Weiss and Tobias 2016, Charbonneau 2020). In 2016, Frank Stefani suggested that the ~11-year solar cycle could be linked to the combined tidal influence of Jupiter, Venus and Earth on the Sun's surface as small tidal changes may tug on the Sun's surficial plasma, affecting surface magnetic activity and the solar wind (Stefani et al 2016).

Many possible ways in which the gravitational influence of the planets might contribute to variability in solar output have been considered. Whichever way it happens, the tidal torque the planets give to the Sun's upper layers is minor in scale. However, it is possible that such processes could be magnified over time if left to run harmoniously for millions of years without disruption (Pikovsky et al 2001).

Solar Activity Archives

Detailed examination of the Sun's surface reveals regions of intense magnetic activity, known as sunspots, in a narrow-latitude band around the solar equator. Sunspots are surficial depressions caused by increased magnetic flux in the convective zone (Solanki 2003). Large magnetic loops generated in and around sunspot zones occasionally crisscross and break, sending bursts of Ultraviolet (UV) radiation and coronal mass ejections outward. Even though sunspots appear darker than the areas around them and their temperature is usually 1000-2000°K cooler than the rest of the Sun's surface, sunspots increase the amount of outgoing UV radiation, leading to a slight increase in the Sun's total energy output.

Sunspot records for the whole Holocene Epoch can be inferred from ^{10}Be and ^{14}C isotope data in ice cores, sediments, stalagmites and tree rings. Both ^{10}Be and ^{14}C are radioactive and originate via the interaction of molecules in Earth's atmosphere and cosmic rays from either the solar wind or the galactic background. Since the galactic background tends to be more constant than variations in solar wind, variations in ^{10}Be and ^{14}C on Earth mostly reflect changes in solar output. Figure 9A provides an overview of the processes involved in transforming these cosmic isotopes into proxies for solar activity. ^{10}Be forms in the atmosphere due to cosmic ray spallation with nitrogen and oxygen molecules (Poluianov et al 2016). ^{10}Be has a long half-life of ~1.39 Myr (Korschinek et al 2010). ^{10}Be atoms eventually precipitate and since they are biologically inert, they accumulate in layers of ice and sediment. ^{14}C forms when free neutrons interact with nitrogen atoms in the upper troposphere. ^{14}C is carried in atmospheric CO_2 molecules and is eventually preserved in biological layers such as tree rings and deep-sea sediment. Note that although ^{14}C has a much shorter half-life of 5730 years, it is much more abundant than ^{10}Be.

Figure 9A: An illustration of how isotopic solar activity proxies are derived. Both ^{10}Be and ^{14}C form due to the interaction between cosmic rays and molecules in the upper atmosphere. ^{10}Be is abiotic and just precipitates and accumulates on surfaces. ^{14}C ends up in CO_2 molecules and integrates with the carbon cycle.

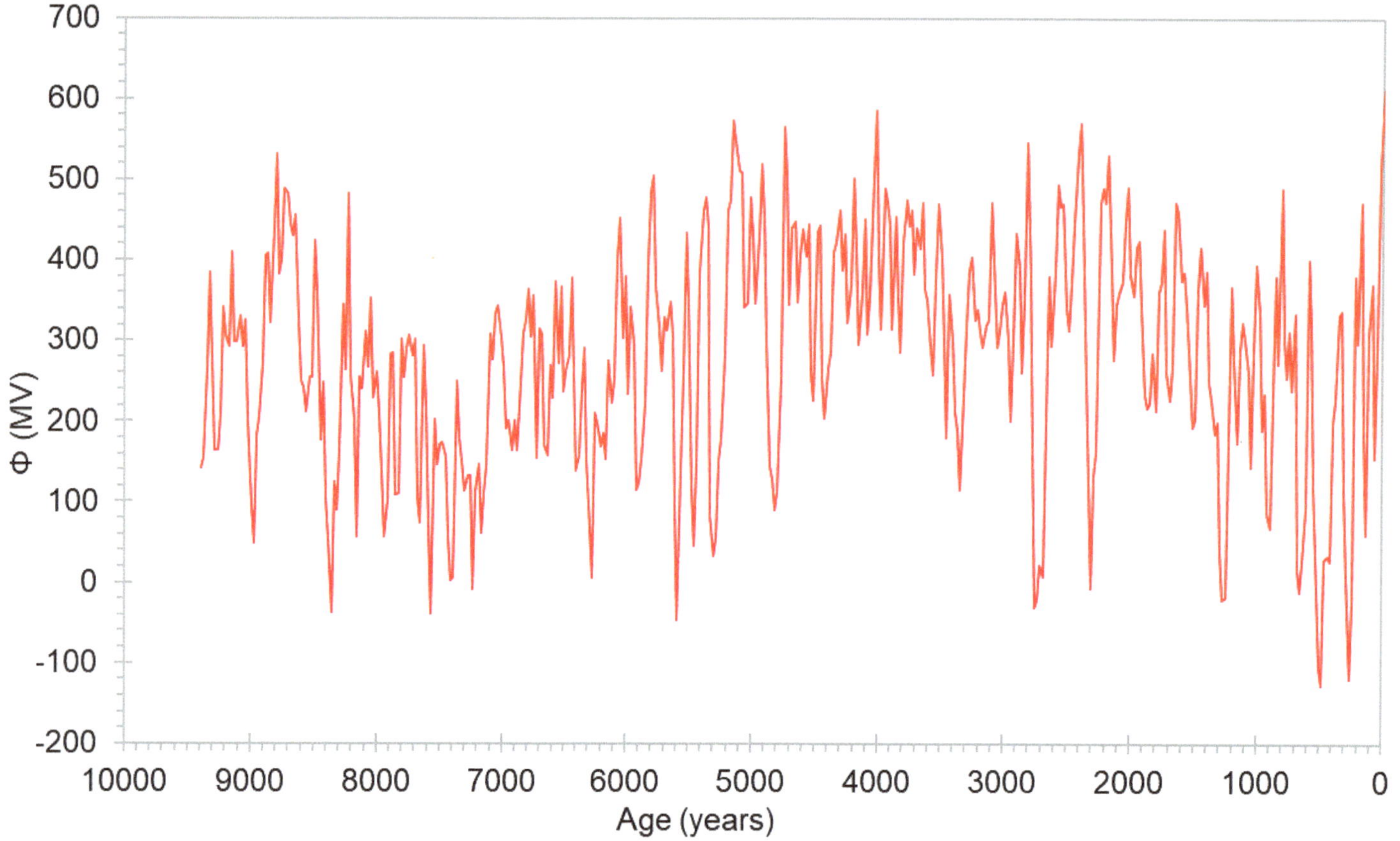

Figure 9B: A combined ^{10}Be and ^{14}C isotope reconstruction of solar activity over the last 9400 years (Steinhilber et al 2012).

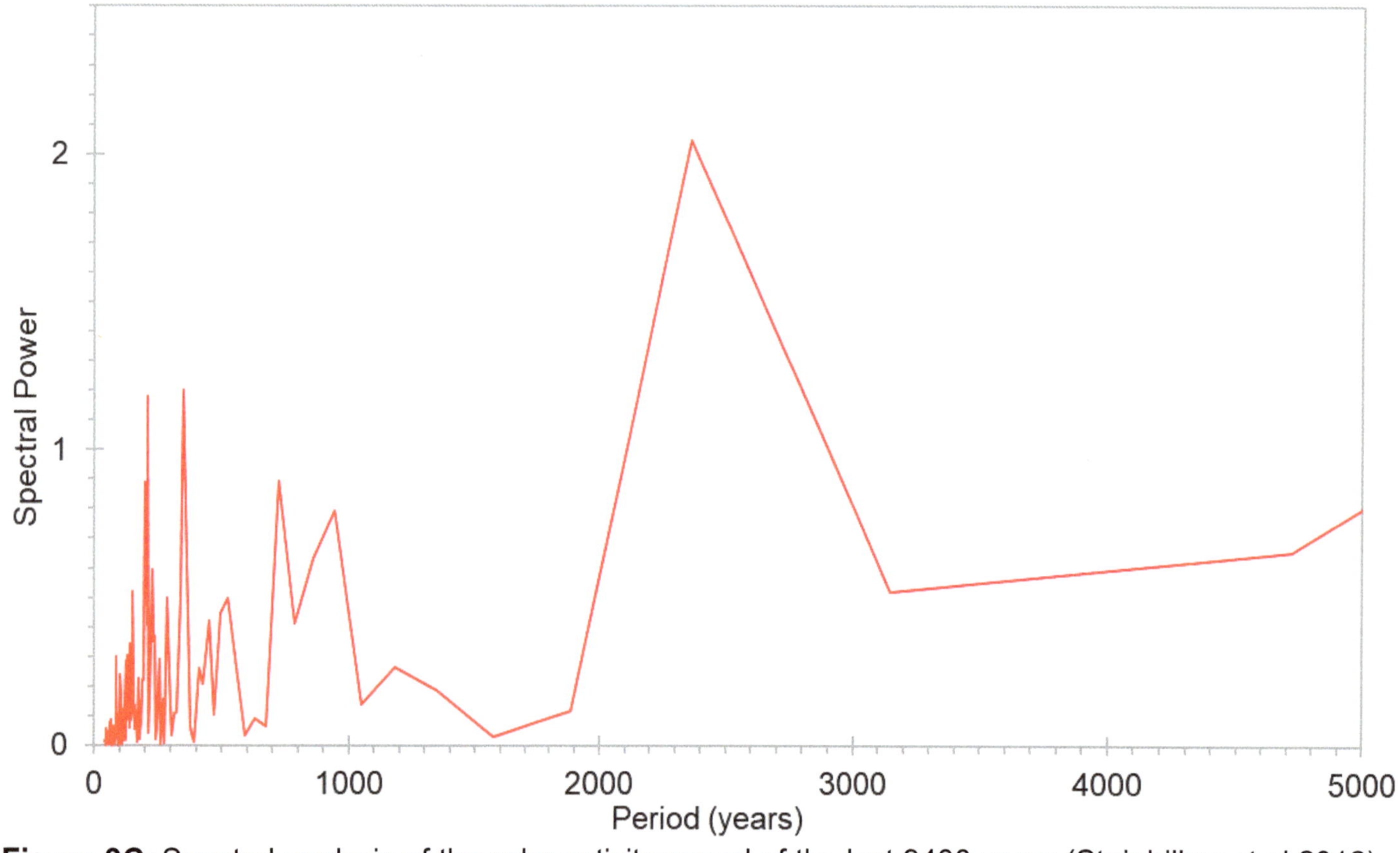

Figure 9C: Spectral analysis of the solar activity record of the last 9400 years (Steinhilber et al 2012).

name	period (years)	Solanki 2004	Steinhilber 2012	Abreu 2012	McCracken 2013	Biswas 2023
Gleissberg	87.3 ± 0.7	3	8	30		18
	104.5 ± 0.6	7		21		
	148 ± 2.0	6	15	42		
Seuss	208 ± 2.4	11	34	58	50	23
	350 ± 8	18	28	55	35	
	420 ± 20	16	5			19
	510 ± 15	29	13	44	33	
	708 ± 28	10	29		43	
Millennial-scale	976 ± 53	21	22		45	30
	1126 ± 71				17	
	1301 ± 96		6		16	
	1768 ± 174				24	
Hallstatt	2310 ± 304	45	51		52	43

Figure 9D: A summary of solar periodicities larger than 11 years commonly reported in scientific literature. The amplitudes are normalized as a percentage of the maximum spectral power of each individual study (Solanki et al 2004, Steinhilber et al 2012, Abreu et al 2012, McCracken et al 2013, Biswas et al 2023).

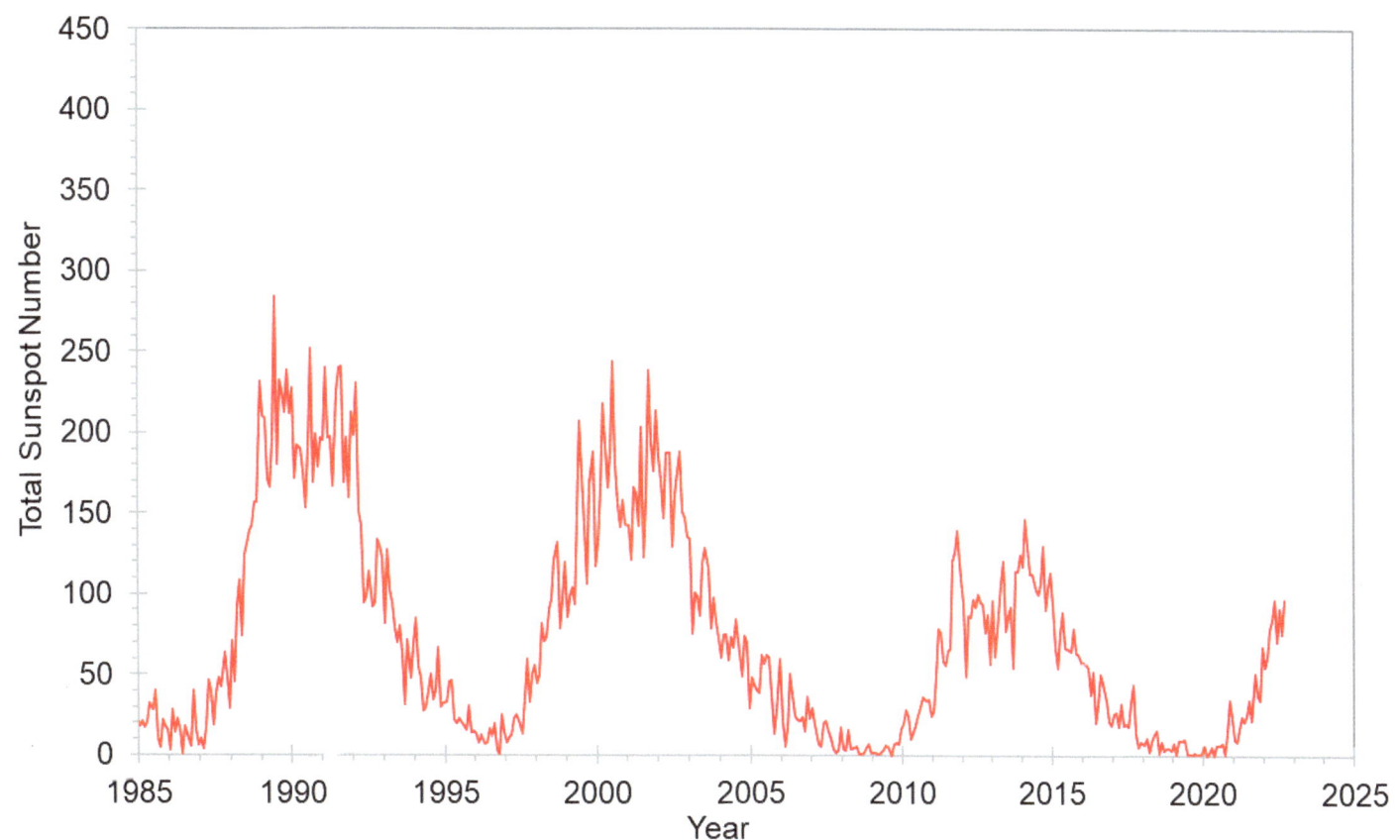

Figure 9E: Solar activity since 1985, showing solar cycles 22, 23 and 24, and the start of solar cycle 25.

Figure 9B shows a multiproxy sunspot number reconstruction compiled by Friedhelm Steinhilber in 2012 using both ^{10}Be and ^{14}C measurements (Steinhilber et al 2012). A spectral analysis of Steinhilber's isotope data is shown in figure 9C, which contains several prominent periodicities near 208, 350, 710, 950 and 2400 years. Figure 9D shows a summary of solar periodicities commonly reported

in isotope studies. Cyclical patterns in the sunspot record vary in amplitude and duration, but the average length of these cycles remains relatively consistent. Figure 9E displays the monthly sunspot number record since the year 1985, showing three full rounds of the basic ~11-year sunspot cycle.

Figure 9F-1: October 15th, 1999, near a Venus-Earth-Jupiter-Saturn tidal maximum.

Figure 9F-2: October 15th, 2010, near a Venus-Earth-Jupiter-Saturn tidal maximum.

Figure 9F-3: July 15th, 2020, near the last Venus-Earth-Jupiter Saturn tidal maximum.

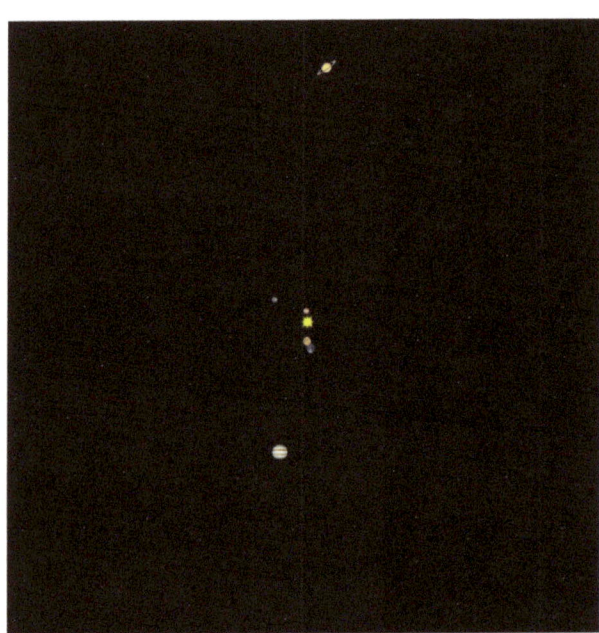

The ~11-year solar cycle

The most prominent sunspot cycle has an average period of ~11 years, but ranges in length from about 9 to 14 years. It's actually a double cycle of ~22 years on average. The Sun's magnetic poles flip halfway through it with its North Pole becoming the South Pole and vice versa. This ~11-year variation in sunspot activity is likely based on the tidal influence of the planets. This is partially deduced by the fact that sunspots occur near the solar equator, where the gravitational influence of planets on the Sun would be strongest. Also of note is that this sunspot cycle's period is close to Jupiter's orbital period of 11.86 years. The last minimum point of the ~11-year cycle happened in early 2020 and the next maximum is expected to occur around the year 2025. As indicated in figure 9E, the previous minimum and maximum before that occurred in late 2008 and early 2014 respectively.

Among the planets, Jupiter's gravitational tidal influence on the Sun is the largest by far, at 11.7 times larger than the value for Earth and about three quarters of the total for all planets combined. The second largest contribution comes from Venus (1.56), followed by Saturn (1.05), Earth (1.00), Mercury (0.37), Mars (0.05), Uranus (0.04), and Neptune (0.02). Jupiter's orbital eccentricity is 0.05, which causes its gravitational influence on the Sun to vary by ~20%. The last time Jupiter reached its perihelion was in February 2011. A particularly large combined torque on the Sun's surface is induced when Venus and Earth line up closely with Jupiter's perihelion. Such a situation occurs every 11.07 years on average (Stefani et al 2016). As you can see in figure 9F to the left, the last three Venus-Earth-Jupiter alignments occurred in October 1999, October 2010 and July 2020. Based on figure 9E's sunspot record, these dates lie near the last three minima of the ~11-year sunspot cycle, within a maximum error bound of about three years.

The ~11-year cycle is also linked to Jupiter-Saturn cycles. Jupiter and Saturn line up about once every 20 years in what some astronomers call a great conjunction. Halfway between each of these alignments, the two planets face exact opposite sides of the Sun. This opposition alignment produces the same tidal effect on the Sun as a conjunction in the same way that the lunar tides produce high tide during both full moon and new moon. Tidal maxima of the Jupiter-Saturn cycle occur with a half synodic period of 9.93 years (Scafetta and Bianchini 2022). The last two great conjunctions were in May 2000 and December 2020, with an opposition occurring in August 2010. These three Jupiter-Saturn alignments roughly coincided with the last three Venus-Earth-Jupiter conjunctions with an error bound of ±8 months.

A combination of Venus-Earth-Jupiter alignments and Jupiter-Saturn conjunctions can roughly explain the general pattern of the ~11-year sunspot cycle over the last two decades, but not the last three decades. As shown in figure 9E, solar cycle 25 began with a solar activity minimum during December 2019. This preceded figure 9F-3's July 2020 Venus-Earth-Jupiter tidal maximum by a matter of months and the December 2020 Jupiter-Saturn tidal maximum by about one year. The previous solar cycle, solar cycle 24, began in December 2008, less than two years before the August 2010 Jupiter-Saturn opposition and the October 2010 Venus-Earth-Jupiter alignment in figure 9F-2. On the other hand, solar cycle 23 began in August 1996, which was more than three years before any planetary tidal maxima.

In 2012, Mikhail Gorbanev of the International Monetary Fund reported a correlation between maxima of the ~11-year sunspot cycle and low unemployment levels in the United States (Gorbanev 2012). According to his study, economic booms tend to occur when the Sun is more active. A quick overview of global economic activity over the last five decades suggests that the last five minima of the ~11-year sunspot cycle have roughly correlated with a series of notable economic crises, with a maximum error bound of ±2 years. March 1976 roughly coincided with the center of the 1970s energy crisis. September 1986 was within a year of 1987's Black Monday. August 1996 preceded the 1997 Asian financial crisis by about one year. December 2008 was not long after the start of the 2008 financial crisis. December 2019 was not long

before the 2020 stock market crash. Although connections between the ~11-year sunspot cycle and economic activity have been disputed for two centuries, the close association between sunspot extrema and notable economic events outlined here is difficult to ignore.

The ~89.5-year solar cycle

In addition to the ~11-year cycle, a number of longer duration solar periodicities are consistently reported in scientific literature. Figure 9D identified prominent solar activity periodicities of 87, 148, 208, 350, 510, 976 and 2310 years. To help visualize how these relate to the ~11-year cycle, figure 9G on the next page shows a longer version of the sunspot record extending back to the first modern sunspot observations in the early 1600s.

A solar cycle with an average period between 80 and 100 years was found in several studies of past solar activity (Ruzmaikin et al 2006). Smaller in amplitude than the ~11-year cycle, this centennial-scale oscillation is often called the Gleissberg cycle. It is highly variable in length but the average period is nearly equal to 3 orbits of Saturn (88.34 years) and 8 rounds of the 11.07 Venus-Earth-Jupiter alignment pattern (88.56 years). In figure 9C, the Gleissberg cycle has a period of ~88.2 years. Some authors describe this cycle as an amplitude modulation of the ~11-year solar cycle (Hathaway 2015). It is not easily detected using spectral or wavelet analysis because it varies so much in length. A better way to see the effect of the Gleissberg cycle on the sunspot record comes from examining the cumulative z-score of monthly sunspot numbers, shown in figure 9H on the next page.

The Gleissberg cycle is likely based on the ~6:15 orbital resonance of Saturn and Jupiter's orbits, which has a regular half period of ~89.5 years. That is, for every 3 orbits of Saturn, Jupiter goes around the Sun ~7.5 times. Figure 9I on page 79 provides an overview of this 6:15 Jupiter-Saturn resonance pattern. The last Jupiter-Saturn alignment in July 2020 was a very close conjunction of all six of the innermost planets. With all of them tugging on the Sun within a small sliver of the sky, the July 2020 conjunction was a time when the combined tidal effect of the planets was maximized. About 90 years before this, in January 1931, a similar arrangement saw the five innermost planets all on one side of the Sun with Saturn near opposition.

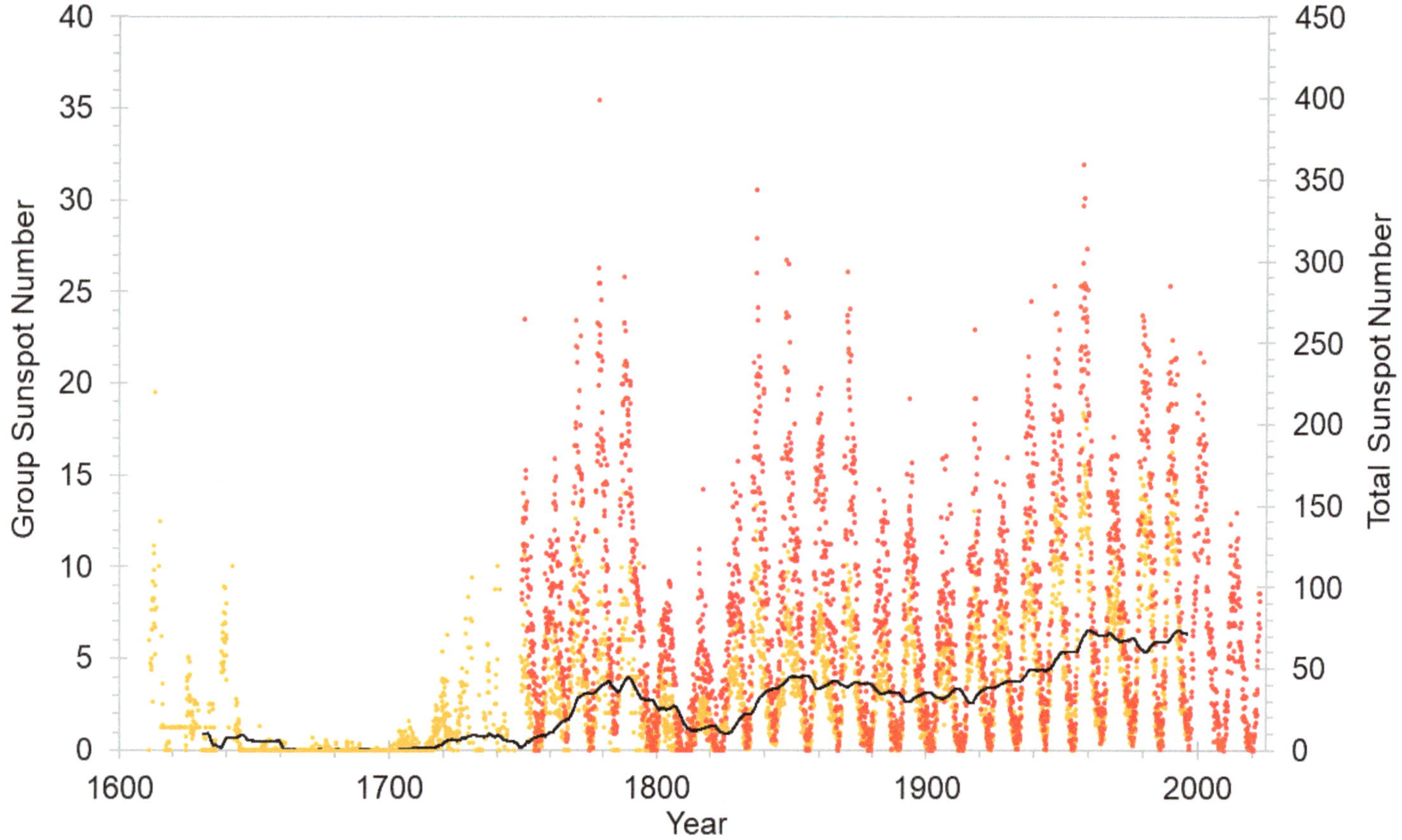

Figure 9G: The sunspot record of the last four centuries. The red dots indicate monthly sunspot counts since 1750 (NOAA 2022). The yellow dots represent group sunspot number counts, which generate a lower resolution approximation of the total sunspot number (Hoyt and Schatten 1998). The black curve is a moving average of the group sunspot number that identifies long term trends in solar variability.

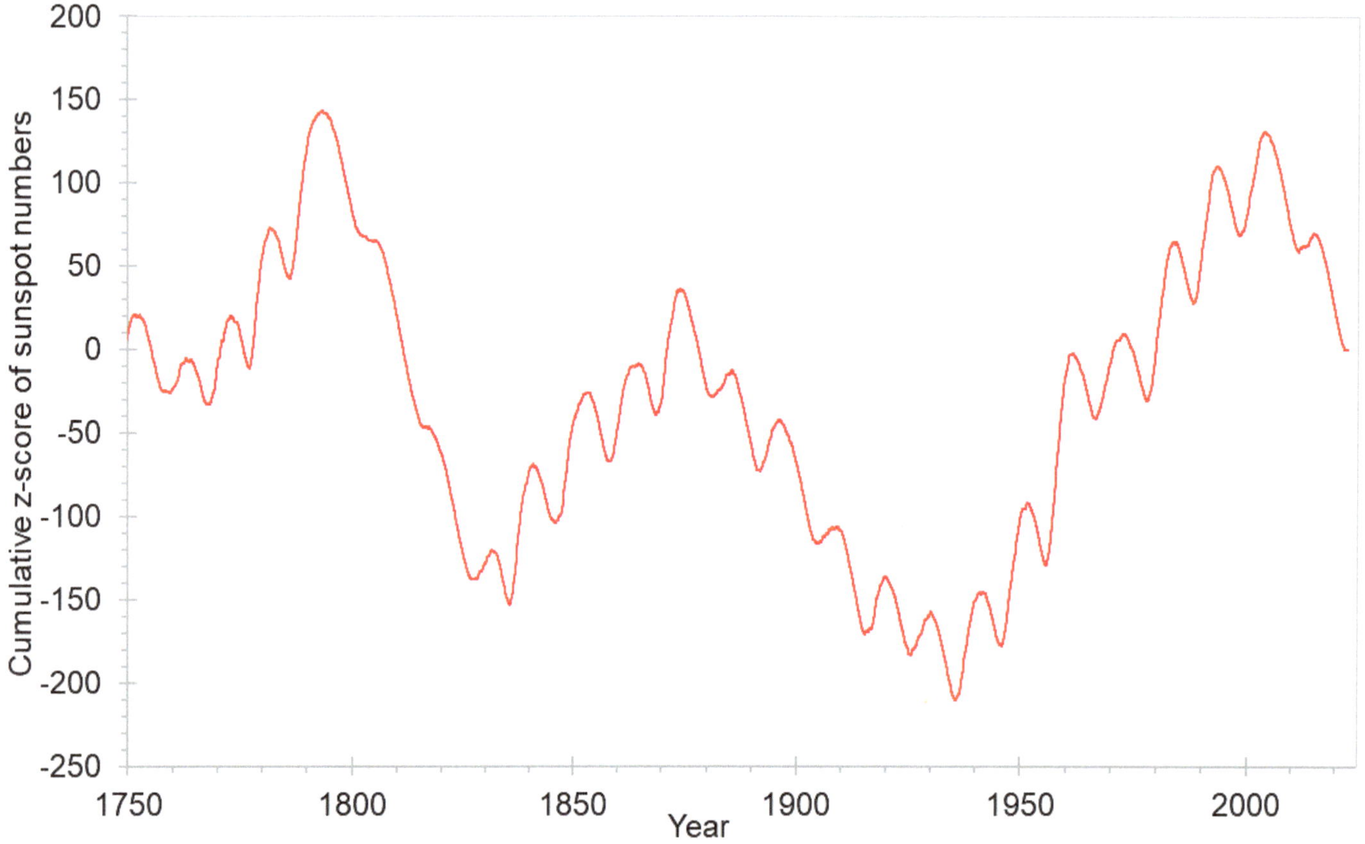

Figure 9H: Cumulative z-score of monthly sunspot counts since 1750 (Wu et al 2018).

Figure 9I-1: November 15th, 1751, near a Venus-Earth-Jupiter-Saturn tidal maximum.

Figure 9I-2: June 15th, 1841, near the Venus-Earth-Jupiter-Saturn tidal maximum.

Figure 9I-3: January 15th, 1931, near the Venus-Earth-Jupiter Saturn tidal maximum.

According to figure 9I's planetary alignments, the last two planetary-induced tidal maxima of the ~89.5-year cycle would have been close to 1931 and 2020. These dates, however, are not in line with most current estimates for the minima of the Gleissberg cycle in scientific literature. One recent study placed the last well-defined minimum of the Gleissberg cycle around the year 1910, while some older studies suggested dates in the 1960s or 1980s (Steinhilber et al 2012, Kopecky 1991, Sonett et al 1991). With this discrepancy in mind, it's important to consider that any repetitive planetary alignment pattern on this timescale could have lagging or lingering effects on solar activity, climate and economic factors, and hence not always line up exactly. The tidal maxima of the Gleissberg cycle may occur somewhat before the peak of its influence on solar activity or climate in a similar way to how many parts of the Northern Hemisphere experience their highest temperatures of the year in early August, rather than in late June when the actual maximum in sunlight occurs. A potential lag in the Gleissberg cycle's causes and effects may also be hinted at by the fact that it shows up clearly in figure 9H's cumulative z-score record yet is next to invisible in the actual sunspot record of figure 9G.

Although the Gleissberg cycle has been described as highly irregular, it isn't unreasonable to suggest that it could be controlled by the Venus-Earth-Jupiter-Saturn alignments in figure 9I, which repeat with a regular period of ~89.5 years. It is interesting that January 1931 and July 2020, the last two tidal maxima of the ~89.5-year cycle hypothesized here, coincide with years near the center of two periods of worldwide economic decline that stand out: the Great Depression from 1929 to 1939 and the recessional period that started in 2008 and continues today. 89.5 years before 1931 was July 1841, which was near the middle of a period of economic depression in the United States between the panic of 1837 and the California Gold Rush of 1848 (Glasner 2013). It was also less than a decade before the Irish Potato Famine of 1845 and the 1848

Revolutions in Europe. 89.5 years before 1841 was January 1752. The 1750s were a time of glacial advance in North America and Iceland but didn't correspond with any well-known historical economic crises, at least not one of exceptional magnitude. Other tidal maxima of the ~89.5-year cycle that occurred during past periods of low sunspot activity include the Maunder Minimum (1645-1715), the Grindelwald Fluctuation (1560-1630), the Sporer Minimum (1450-1550) and the Wolf Minimum (1280-1350)(Eddy 1976, Usos-kin et al 2015, Jones et al 2021, Kotze 2023). A summary of these types of tidal maxima and corresponding historical events is shown in figure 9J.

date	Saturn in...	associated economic, climatic or solar activity interludes
Jul 2020	conjunction	Great Recession (2007-2009), COVID-19 Recession (2020)
Jan 1931	opposition	Great Depression (1929-1939)
Jul 1841	conjunction	Panic of 1837, California Gold Rush (1848-1855)
Jan 1752	opposition	
Jul 1662	conjunction	Maunder Minimum (1645-1715)
Jan 1573	opposition	Grindelwald Fluctuation (1560-1630)
Jul 1483	conjunction	Sporer Minimum (1450-1550)
Jan 1394	opposition	
Jul 1304	conjunction	Wolf Minimum (1280-1350)

Figure 9J: Jupiter-Saturn tidal maxima and contemporary periods of low economic activity.

date	associated economic events	start	end	middle	Δ
Oct 1975	3rd industrial revolution (information age)	1950	2000	1975	-0.75
Apr 1886	2nd industrial revolution	1871	1914	1892.5	+6.25
Oct 1796	1st industrial revolution	1760	1840	1800	+3.25

Figure 9K: Midway points between two of figure 9I's Jupiter-Saturn tidal maxima pattern alongside periods of enhanced economic activity.

date	associated political events	start	end	middle	Δ
May 1953	Civil Rights Movement	May 1954	Aug 1968	Jun 1961	+8.08
Nov 1863	American Civil War	Apr 1861	May 1865	Apr 1863	-0.58
May 1774	American Revolutionary War	Apr 1775	Sep 1783	Jun 1779	+5.08

Figure 9L: Center of the waning parts of figure 9I's Jupiter-Saturn tidal maxima pattern compared to important American internal conflicts over the last few centuries.

date	associated economic events	date	Δ
Feb 1998	Euro dollar introduced	Jan 1999	+0.92
Aug 1908	US dollar became an important reserve currency after World War I	Nov 1918	+10.25
Feb 1819	British pound sterling became an important reserve currency	Nov 1815	-3.25

Figure 9M: Center of the waxing parts of figure 9I's Jupiter-Saturn tidal maxima pattern and important events related to global reserve currencies.

For this hypothetical ~89.5-year oscillation in solar activity, the halfway point between two tidal maxima could resemble a tidal minimum, but it actually represents more of a return to a background planetary tidal equilibrium with lengthy periods of non-special alignments. The four most recent of these midpoints would have occurred in October 1975, April 1886, October 1796 and April 1707. Since tidal maxima of the Gleissberg cycle seem to correspond with periods of reduced economic activity, it's intuitive to suggest that perhaps these midway points could correspond with times of heightened economic activity. Figure 9K shows how the expected dates for the midpoints of the ~89.5-year cycle relate to the beginning, middle and end of the first, second and third industrial revolutions. These were times of notably high levels of technological innovation and economic productivity. The first industrial revolution witnessed a dramatic increase in textile manufacturing and steam power generation between 1760 and 1840 (Crafts 1994). The second industrial revolution was characterized by a rapid increase in steel manufacturing and the expansion of railway and electricity networks between 1871 and 1914 (Mokyr and Strotz 1998). The third industrial revolution spanned the last half of the twentieth century and saw a great expansion in the use of computers and digital information (Mohajan 2021). Based on this pattern, a fourth industrial revolution can be predicted with a center close to April 2065. As you can see in figure 9K, the centers of the three industrial revolution only diverge from the midpoints of the ~89.5-year cycle by a maximum of ±6.25 years.

For the hypothetical ~89.5-year cycle outlined here, a quarter period after a tidal maxima might represent a time when the planetary tidal influence on the Sun on centennial timescales is waning the fastest. The central part of this waning phase would act as a parallel to the spring equinox and correspond with times when economic activity would be increasing the fastest. According to the pattern in figure 9I, the last three of these central waning events would have occurred in May 1774, November 1863 and May 1953. As you can see in figure 9L, these three dates were near the midpoint of three important internal conflicts in American history that stand out: the American Revolutionary War, the American Civil War and the Civil Rights Movement. Based on this pattern, the next alignment of this type should occur in November 2042.

A quarter cycle before each tidal maximum of the ~89.5-year cycle would represent times when the planetary tidal influence on the Sun on a centennial timescale is increasing the fastest. These events might act as parallels to the autumnal equinox and correspond with periods when economic activity, at least on the ~89.5-year timescale, would be declining the fastest. According to the pattern in figure 9I, the last three of these events would have occurred in February 1819, August 1908 and February 1998. As you can see in figure 9M, these three dates have been close to some historically significant events in global finance when a new important reserve currency was introduced. Based on this pattern, the next alignment of this type is expected to occur in August 2087.

The analysis performed here has shown that economic and political events associated with the ~89.5-year cycle are well pronounced even though the Gleissberg cycle is generally not seen in isotope proxies as being very large in amplitude. The cycle's average length and its potential effects on human societies bears eery resemblance to sociological cycles identified in Strauss-Howe Generational Theory (Strauss and Howe 1991). Both Strauss-Howe generational cycles and Gleissberg oscillations vary significantly in length but on average have a period between 80 and 100 years. The historical events highlighted in figures 9J, 9K, 9L and 9M don't line up precisely with the events outlined by Strauss and Howe but follow a somewhat similar progression. Therefore, it is somewhat difficult to tell whether the patterns described here are driven primarily by solar variation or by inherent sociological factors. An 80 to 100-year cycle covers about one human lifespan and about four human generations so it is possible that cyclical patterns on this timescale reflect a progression of economic and political events related to one human generation's response to the actions of the previous generation. That being said, there is a possibility that the planetary alignments shown in figure 9I have an influence on the historical events outlined in figures 9J, 9K, 9L and 9M as a maximum difference of only ±10.25 years exists between the midpoint of these historical events and their corresponding planetary alignments. This represents a maximum error bound of ~11.5%, which is much narrower than one would expect from randomly distributed dates.

The ~170-year solar cycle

Another commonly identified solar cycle has a period close to 170 years, although in figure 9D the closest periodicity to this is 148+2.0 years. Because a period of 170 years matches better with historically significant climatic events on this timescale, it will be referred to as the ~170-year cycle in this article. Variations in solar activity on this approximate timescale could be linked to the beating among harmonics of Jupiter and Saturn with the ~11-year solar cycle (Abreu et al 2012, Scafetta 2014). Although this solar periodicity is relatively low in amplitude compared to others, three important climate-related events in the historical record may be linked to it. The last three tidal maxima of this cycle probably occurred around the years 1640, 1810 and 1980. In figure 9G, the black curve representing the moving average of monthly sunspot numbers has notable dips around these dates.

The most recent tidal maximum of this ~170-year solar cycle would have occurred around 1980 and corresponded with an interesting time in history. In the 1960s, authors like J. Murray Mitchell and Cesare Emiliani were warning about the dangers of global cooling due to a gentle but persistent decline in global temperature (Mitchell 1963). These concerns were echoed in Time and Newsweek magazines in the 1970s. Concerns about global cooling were then replaced with concerns about global warming in the 1980s (Conway 2008). Thus, the year 1980 signals an important event in recent climatic history: the transition from global cooling to global warming.

About 170 years before 1980 was the year 1810. This was the very center of the Dalton Minimum, a period of exceptionally low sunspot activity between 1780 and 1840 (Anet et al 2014). The so-called "year without a summer", 1816, when crop failures and famines plagued North America and China, also falls within this interval (Schurer et al 2019). However, the abnormally cool conditions associated with 1816 are usually attributed to the eruption of Indonesia's Mount Tambora the previous year rather than being based primarily on solar variability (Wagner and Zorita 2005). Nonetheless, the Dalton minimum as a whole was noticeably colder and witnessed lower solar activity than the decades before and after it.

About 170 years before 1810 was the year 1640. That was close to the beginning of a lengthy period of extremely low sunspot activity between 1645 and 1715 known as the Maunder Minimum (Usoskin et al 2015). This period was characterized by cold temperatures, crop failures and economic hardship in Europe, North America, East Asia and elsewhere (Lamb 1977). The year 1640 precedes the beginning of the Maunder Minimum by five years. This minimum of the ~170-year cycle could have contributed to low solar activity during the mid-17th century, but it is just one of many solar periodicities whose minimum points occurred around this time.

The last three tidal minima of this hypothetical ~170-year cycle would have occurred around the years 1555, 1725 and 1895. The tidal minimum around 1725 may have worked to counteract an expected tidal maximum of the ~89.5-year cycle around 1752. The last tidal minimum of the ~170-year cycle around 1895 may have enhanced the ~89.5-year cycle's tidal minimum of 1886. Based on this pattern, the next tidal minimum of the ~170-year solar cycle could occur around the year 2065.

The ~208-year solar cycle

Another important solar cycle with a period of ~208 years is called the Seuss cycle by some and the de Vries cycle by others (Biswas et al 2023). It was identified in all five isotope studies listed in figure 9D. In figure 9C's spectral analysis it has a period of ~209.7 years. The last tidal maximum of the Seuss cycle occurred somewhere between 1888 and 1898, near tidal minima of both the ~89.5-year cycle and the ~170-year cycle (Sonett et al 1991, Breitenmoser et al 2012). This overlaps the last part of the Long Depression, an economic recession in Great Britain and the United States from 1873 to 1896 (Fletcher 2013). A previous tidal maximum of the Seuss cycle would have occurred around the year 1685, near the center of the Maunder Minimum. 208 years before that, a Seuss tidal maximum would have occurred around 1477 and coincided with the early part of the Sporer Minimum, a period of low sunspot activity between 1460 and 1555 (Eddy 1976).

The ~208-year cycle's effect on the sunspot record may be seen in figure 9G as two bowl-like depressions in the moving average centered around its last two tidal maxima, 1685 and 1893. Based on this pattern, the last three tidal minima of the Seuss cycle would have been near 1576, 1784 and 1992. These dates correspond with higher-than-average solar activity but do not stand out as very significant in a historical sense. The next tidal maximum of the ~208-year cycle can be expected to occur around the year 2100.

The ~208-year cycle shows some dissimilarities to shorter-period solar cycles. It tends to be more pronounced in times of low overall solar activity and less visible in the record during times of moderate overall solar activity (Biswas et al 2023). This means that it behaves less sinusoidally and more like a repeating negative delta function. This can explain why its tidal minima don't have a noticeable effect on solar activity, climate or economic history while its tidal maxima do. The cause of the ~208-year cycle appears to be based less on planetary alignments and more on dynamics of the Moon-Earth-Sun system. 208 years equals roughly 19 rounds of the ~11-year solar cycle and 11 rounds of the Metonic cycle. The Metonic cycle is an 18.6-year nutation period of the Moon with respect to the ecliptic, a recurring pattern where every ~19 years, the same phase of the Moon occurs on roughly the same day of the calendar year. The Metonic cycle is responsible for secondary lunar tidal oscillations that produce high and low astronomical tides on Earth. Even more notable is the fact that every ~208 years, the Moon is in the same position of the sky with respect to the Sun and stars. This basically means that the Moon-Earth-Sun-Stars astronomical clock resets itself with respect to Earth's orbit and the stars every ~208 years. Although the Sun has a larger effect on Earth's climate, the Moon's influence should not be underestimated. It is the largest satellite in the inner solar system. With a mass about one fifth that of Mercury, its tidal effect on the Sun is ~1% of the Earth's.

The ~350, ~500 and ~708-year solar cycles

Another important solar cycle has a period of ~350 years. Mathematician Valentina Zharkova refers to it as the grand cycle and predicts that its next solar minimum around the year 2036 will cause a significant reduction in Earth's temperature (Zharkova et al 2015, Zharkova 2020). This ~350-year oscillation in solar activity is likely related to a beating effect produced by the superposition of two shorter period waves, both close to 22 years in length and based in different layers of the Sun's interior (Zharkova et al 2023). The last tidal maximum of the ~350-year solar cycle would have occurred around 1686, near the middle of the Maunder Minimum and nearly overlapping tidal maxima of the ~89.5-year, ~170-year and ~208-year solar cycles. The ~350-year cycle would have also had a tidal maximum around 1336 during the Wolf minimum, a period of low sunspot activity between 1280 and 1350 (Zharkova et al 2018).

J.A. Abreu's research team detected a significant solar periodicity at ~508 years (Abreu et al 2012). Several groups of Chinese scientists have also reported a strong periodicity close to 500 years, with the last solar maximum occurring around the year 1900 (Xu et al 2014, Ma et al 2018). Although this ~500-year cycle is consistently detected in isotope studies, it varies greatly in relative magnitude. It is sometimes referred to as a dominant periodicity but is sometimes barely discernable from the statistical background.

Another periodicity of ~708 years was detected in Ken McCracken's 2013 isotope study but is not consistently reported as a prominent solar variability pattern (McCracken et al 2013, Usoskin 2017).

The millennial scale solar cycle

A millennial-scale periodicity in solar activity is consistently identified in isotope studies. Often called the Eddy cycle, its period usually ranges between 900 and 1000 years. In figure 9C, it is represented by a broad peak centered at ~944 years. A sizable body of evidence suggests that it may be responsible for major changes in climate during the Holocene (Wickson 2023). On a century-to-century basis, the millennial-scale solar cycle's influence on climate can be seen in the oscillation from the Roman Warm Period (250 BC-400) to the Late Antique Little Ice Age (536-660) to the Medieval Warm Period (950-1250) to the Little Ice Age (1300-1850) and finally to the Modern Warm Period (Campbell et al 1998, Matthews and Briffa 2005, Mann et al 2009, Buntgen et al 2016).

In January 1486, during the middle of the Little Ice Age, a rare planetary alignment occurred in which all four giant planets lined up within a ~30° sector of the sky. This alignment, shown in figure 9N-1, marks the last tidal maximum of the millennial-scale solar cycle. Despite the fact that Uranus and Neptune have smaller tidal impacts on the Sun than Mars, the fact that this rare alignment happened in the middle of the Little Ice Age is notable.

Figure 9N-1: The positions of the outer planets on January 15th, 1486.

A similar tidal maximum involving the outer planets occurred in July 590. At this time, Jupiter, Saturn and Neptune were within a ~5° sector of the sky, and all four outer planets were contained in the same quadrant, as illustrated in figure 9N-2. This planetary alignment corresponded with a sunspot and temperature minimum during what's called the Late Antique Little Ice Age. This was during the European Dark Ages when the city of Rome reached a low point in population.

The two planetary alignments described here suggest that tidal maxima of the millennial-scale solar cycle occur during Jupiter-Saturn-Neptune conjunctions where Uranus is not in opposition. Using the examples shown in figures 9N-1 and 9N-2, a set of approximate dates for the expected maxima and minima of the millennial-scale solar cycle can be generated. Although, using this pattern, the cycle would have an actual period of ~895.5 years. This is slightly below the lower bound of the usual range reported in isotope studies, but it is close to the amount of time it takes the aphelia of Jupiter and Saturn to line up in the same direction (Murray and Dermott 1999). Based on this pattern, the next tidal maximum of the millennial-scale solar cycle would be expected in July 2381. The last three tidal minima of this cycle probably occurred close to October 142, April 1038 and October 1933.

Figure 9N-2: The positions of the outer planets on July 15th, 590.

The millennial-scale solar cycle is among the strongest of solar periodicities in figure 9D and in terms of its association and progression with historical events. Its maxima and minima line up with the historically significant climate periods outlined in figure 9O with a maximum error bound of ±89 years, or about 10% of the cycle's length. Like the Gleissberg cycle, some historical patterns attributed to the millennial-scale solar cycle might have a component related to ecological and sociological dynamics rather than just planetary alignments. A summary of how different parts of the millennial-scale cycle may be reflected in the historical record is as follows.

Millennial tidal maxima, the winter part of this cycle, tend to be associated with population decline in cities. They also tend to favour the development of city states and small regional polities rather than large, centrally governed empires. An example of this is found in comparing the city states of Renaissance Italy during the Little Ice Age to Italy during the Roman Warm Period when the entire Mediterranean region was ruled by one centrally administered state. Millennial winters have also tended to coincide with the introduction of new religious belief systems, such as the beginning of the Islamic religion near the middle of the Late Antique Little Ice Age and the Protestant Reformation in the 16th century. They also tend to be characterized by southward shifts in power and population.

The spring phase of the millennial-scale solar cycle represents the middle parts of transitions from tidal maxima to tidal minima. Millennial springs are often characterized by significant expansions in interregional trade, such as the introduction of new maritime trade routes during the Viking Era and the rapid growth of oceanic trade routes in the 18th century. There tends to be a general shift in power and population towards higher latitudes and coastal regions during millennial springs. At the same time, intrastate conflicts including revolutions, revolts and struggles between social classes tend to be more common.

Tidal minima of the millennial-scale solar cycle correspond with the centers of its summer phases. During millennial summers, urban populations tend to increase and large, centrally governed empires tend to be more common. Millennial summers also tend to overlap historically significant high points in literature, art, architecture and luxury.

date	type	associated historical events	start	end	middle	Δ
Oct 1933	tidal minimum	Modern Warm Period	1850			
Jan 1486	tidal maximum	Little Ice Age	1300	1850	1575	+89.0
Apr 1038	tidal minimum	Medieval Warm Period	950	1250	1100	+61.8
Jul 590	tidal maximum	Late Antique Little Ice Age	536	660	598	+7.5
Oct 142	tidal minimum	Roman Warm Period	250 BC	400	75	-67.8

Figure 9O: Predicted tidal minima and maxima of the millennial-scale solar cycle based on conjunctions of the outer planets compared to dates of historically significant climatic intervals.

Autumn phases of the millennial-scale solar cycle represent the middle parts of transitions from tidal minima to tidal maxima and tend to coincide with periods of scarcity and migration. Millennial autumns tend to induce shifts in power and population away from coasts and towards continental interiors, like what was seen during the Mongol Era. They also tend to promote the development and spread of military technology.

A more detailed analysis of how the millennial-scale solar cycle may have affected human societies during the Holocene is available in a recent book entitled "The 900-Year Climate Cycle" (Wickson 2023). Although the millennial-scale solar cycle is lower in amplitude than the ~2400, ~350 and ~208-year cycles in figures 9C and 9D, it seems to resonate well with macroscopic changes in human societies. As it seems to be based on the arrangement of the four outer planets, it may have a highly variable period between individual cycles that makes it harder to detect using spectral analysis. It

also seems to be occasionally obscured by destructive interference from larger-scale and smaller-scale periodicities.

The ~2400-year solar cycle

The largest solar periodicity consistently detected in Holocene isotope studies, in both period and amplitude, is the Hallstatt cycle. In figure 9D, its period is listed as 2310±304 years. Although other periodicities with longer timescales have been identified in some isotope studies, they are far less significant and consistent. The Hallstatt cycle is linked to a ~2400-year cycle based on close conjunctions of the four outer planets, as illustrated in figure 9P. These conjunctions generate a repeating ~2400-year pattern in the Sun's position relative to the solar system's barycenter (Palus 2007, Biswas 2023).

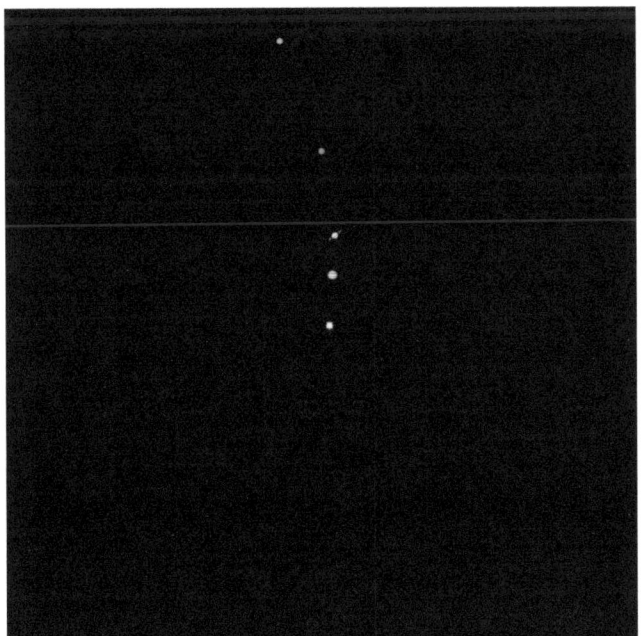

Figure 9P-1: The positions of the outer planets on January 15th, 3322 BC.

Figure 9P-2: The positions of the outer planets on June 15th, 917 BC.

date	associated cool period	Δ	associated historical events	Δ
1486	Little Ice Age (1300-1850)	+89	Renaissance (1400-1600)	+14
917 BC	Hallstattzeit cold epoch (750-400 BC)	+341	Axial age (800-200 BC)	+416
3322 BC	Uruk cool period (3200-2800 BC)	+321	Earliest writing (3400-3100 BC)	+71

Figure 9Q: Comparison between predicted dates for tidal maxima of the ~2400-year solar cycle and three historically significant cool periods.

Although the period of the Hallstatt cycle has varied from ~2100 to ~2500 years over the course of the Holocene, it fits a period of 2402±14 years in the overall ^{14}C record of the last 12.5 kyr (Scafetta et al 2016). One ^{14}C maximum of the Hallstatt cycle is well-known among dendrochronologists and archaeologists, who describe a period from 800 BC to 400 BC where radiocarbon measurements are unreliable due to an issue with the radiocarbon calibration curve (James and Thorpe 1991, Van der Plicht 2005, Fahrni et al 2020). As ^{14}C ratios in that time interval are so much higher than the background, the calibration curve no longer follows an exponential decay model, and hence can't accurately resolve radiocarbon dates.

High ^{14}C levels on Earth occur during times of decreased solar activity because a quieter sun allows more cosmic rays from the galactic background to make it into the Earth's atmosphere. (Stuiver and Quay 1980) This is because a weaker solar wind shrinks the Sun's magnetic blanket around the solar system, which decreases its ability to deflect incoming galactic cosmic rays.

The last Hallstatt tidal maximum occurred during the depths of the Little Ice Age, sometime between 1400 and 1645 (Sonett et al 1991, Scafetta et al 2016). The model presented in this article will place the last Hallstatt solar activity minimum in January 1486, which is also the last tidal maximum of the millennial-scale solar cycle. 1486 was more than a century before the most intense part of the Little Ice Age in the 17th century. The reason for this phase shift is unknown. It could represent a lag between planetary tidal influence on the Sun and its effect on Earth's climate system. It could also be the case that the Maunder Minimum, the most intense part of the Little Ice Age, is based more on the superposition of minima of several smaller-scale solar periodicities rather than being primarily based on the Hallstatt and millennial-scale cycles.

A previous tidal maximum of the Hallstatt cycle occurred sometime between 850 and 600 BC (Martin-Puertas et al 2012). In fact, the Hallstatt cycle's name comes from the Hallstatt culture, who had settlements in Central Europe during this part of the

Late Bronze Age and Early Iron Age. This interval also overlapped the Greek Dark Ages, a period of unusually high cosmogenic ^{14}C concentrations, and a cold period in the Northern Hemisphere inferred from a variety of sources (Davis et al 1992, Martin-Puertas et al 2012, Kronig et al 2018, Robles et al 2022). However, based on the planetary alignment patterns in figure 9P, this article's model of solar periodicities will use a date of 917 BC for the Hallstatt cycle's second last tidal maximum, almost two centuries before commonly reported estimates for the Hallstattzeit cold epoch (750-400 BC)(Damon and Sonett 1991). An extreme high in planetary tidal influence on the Sun occurred around this date that led to an extreme low in sunspot activity, which in turn reduced the solar wind and maximized the flux of ^{14}C from galactic cosmic rays into Earth's atmosphere. After 917 BC, it may have taken several centuries for ^{14}C concentrations on Earth to return to normal levels. The 917 BC Hallstatt tidal maximum predated a tidal minimum of the millennial-scale solar cycle by less than two centuries, which is less than a quarter of the millennial-scale solar cycle. This could have resulted in a large degree of destructive interference between the two cycles, which would explain why the millennial-scale solar cycle has often avoided detection in isotope studies and has not been discussed at length in academic literature.

About 2400 years before 917 BC was 3322 BC, during the late part of the Uruk period. This was about a century before a cool period identified by some paleoclimatologists between 3200 and 2800 BC (Damon and Sonett 1991). With this in mind, a link between the last three Hallstatt tidal maxima and three historical cool periods can be tested. Figure 9Q shows that the centers of the Uruk cool period (3200-2800 BC), the Hallstattzeit cold epoch (750-400 BC) and the Little Ice Age (1300-1850) followed the last three tidal maxima of the Hallstatt cycle with a consistent lag of 215±126 years. These three dates also seem to correspond with three highly significant periods in history that stand out as times of dramatic societal and philosophical transformation: the introduction of writing in Egypt, Mesopotamia and the Indus Valley in the late 4th millennium BC, the Axial Age of the 8th to 3rd centuries BC, and the Renaissance of the 15th and 16th centuries. Figure 9Q also summarizes the relationship between these three historical periods and predicted tidal maxima of the Hallstatt cycle. As these three historical eras each spanned more than a century, they could be related to underlying long-term trends of the Hallstatt oscillation rather than decadal-scale lows in solar activity and/or climate. Based on the pattern outlined here, Hallstatt tidal minima would have occurred near 285 AD, 2117 BC and 4519 BC. Yet these dates do not correspond with any notable repeating pattern in the historical record. The next maximum of the ~2400-year solar cycle could occur around the year 2687.

The relationship between the Hallstatt cycle and the millennial-scale solar cycle is peculiar in that even though the Hallstatt cycle is larger in amplitude, its effect on human societies seems secondary to the effects of the millennial-scale solar cycle most of the time. Only the tidal maxima of the Hallstatt cycle seem to have a strong effect on repeating historical events whereas every quarter-phase of the millennial-scale solar cycle corresponds with a noticeable effect on human societies. For this reason, the Hallstatt cycle's effect on climate and human societies exhibits behaviour more akin to a repeating negative delta function than a sinusoidal pattern.

The Hallstatt cycle's prominence is countered by chaotic background signals in the solar activity record. In that sense, it can be described as a probabilistic pattern that repeats with a varying period. This essentially means that the tidal maxima of the Hallstatt cycle are not perfectly timed, yet there is a tendency for low sunspot activity to occur around the Hallstatt cycle's tidal maxima. Although some intervals of low sunspot activity occur outside of the cool phases of the Hallstatt cycle, about half of the periods of low sunspot activity in the Holocene have occurred within ±250 years of a Hallstatt solar minimum (Usoskin et al 2016). Since the Hallstatt cycle has been shown to behave in this way, it might be instrumental in revealing a deeper understanding of how smaller-scale solar periodicities operate and relate to patterns in climate and human civilizations.

A Composite Model of Solar Variability

Presented here is a model of long-term solar variability based on the superposition of periodic negative delta functions. Solar cycles are quasi-periodic, meaning that each individual round of a cycle is not exactly the same length. In terms of solar output, many solar periodicities tend to be more pronounced around their tidal maxima and less noticeable in other parts of the cycle. Therefore, it is not entirely appropriate to use sinusoidal functions to model them. Instead, repetitive beats of planetary tidal torques result in time-varying solar output patterns that can be better approximated by series of periodically placed pulses of the form

$$S_P(t) = -\sum_{i=1}^{n} A_i e^{-\left(\frac{t-\phi_i}{B_i}\right)^2}$$

Here, $S_P(t)$ represents a particular pattern in solar activity such as the ~11-year solar cycle, $S_{11}(t)$ or the ~208-year solar cycle, $S_{208}(t)$. A_i is the amplitude of each individual pulse of a given solar cycle and B_i reflects the breadth of the Gaussian-shaped dip in solar activity. ϕ_i is the phase shift related to the pulse of maximal tidal torque and minimal sunspot activity. If the breadth of a periodic delta function exceeds one third of its period, it essentially approximates as a sinusoid.

In order to generate a composite model of solar periodicities, an average amplitude, breadth and period will be used to describe each of the component cycles. This has the effect of turning the coefficients $A_1, A_2, A_3, \ldots$ into standard constants for each solar periodicity that can be deduced from figure 9D or historical indicators. In this model, each solar periodicity is described as a set of solar activity minima generated by $\phi_i = \phi_0 - iP$, where ϕ_0 is the most recent sunspot minimum, P is the cycle's period, and i is an integer counter. Describing solar variation in this way allows one to create a relatively simple mathematical model for sunspot cycles that can be easily compared with trends in the historical sunspot record. This model will only focus on solar periodicities longer than 11-years and will be statistically compared with the multi-decadal moving average of the four-century-long historical sunspot record shown in figure 9G. Although actual sunspot cycles do not show the perfect symmetry that these Gaussian-shaped mathematical forms exhibit, this composite model can be used as a preliminary test to check whether solar variation may be related to planetary alignments.

The repeating patterns in this model reflect complicated dynamics of planetary tidal torque oscillations. Jupiter is the main facilitator and the strength of the pull in its direction beats on a range of timescales based on the influence of the other planets. When the combined planetary tidal influence is maximized, the number of sunspots decreases and solar output declines. When these tidal maxima are over, the Sun's output gradually returns to more of an equilibrium position. The sunspot maxima of individual periodicities merely represent the midpoints between two sunspot minima. In other words, planetary tidal periodicities only tend to induce sunspot minima and sunspot maxima manifest when the tidal influence is removed. That being said, some solar periodicities produce wider sunspot minima which makes them appear more sinusoidal in nature, while others have more concentrated sunspot minima that don't last and manifest more like periodic spiky pulses. For example, figure 9E shows the ~11-year solar cycle approximating a sinusoid whereas the ~350-year solar cycle tends to be only identifiable near its tidal maxima. Based on the overview of solar periodicities provided in this article, it seems possible that periodicities that are primarily based on the perihelia of Jupiter and Saturn tend to act more sinusoidal in nature whereas those that aren't tend to produce narrower sunspot minima. This characteristic is exhibited in the significant contrast between the Hallstatt cycle and the millennial-scale solar cycle. The millennial-scale solar cycle, although not always dominant, manifests in a very sinusoidal way with the four different phases of the cycle, its minima, waxings, maxima and wanings, each showing an effect on climate and human societies. Conversely, the Hallstatt cycle acts more like a repeating Gaussian pulse. For the Hallstatt cycle, expected sunspot maxima midway between two consecutive sunspot

minima are unpronounced and simply belong to a lengthy pseudo-equilibrium period. That being said, Hallstatt sunspot minima are so strong that they dominate even if they coincide with the maxima of other periodicities.

The model presented here is composed of six commonly reported solar periodicities larger than 11 years, summarized in figure 9R. A graphical representation of this model of long-term solar activity is displayed in figure 9S alongside the moving average of group sunspot numbers from figure 9G. The two curves in figure 9S have a Pearson correlation coefficient of 93% in the interval from 1705 to 1995.

The Maunder Minimum (1645-1715) acts as a common low point for the ~89.5-year, ~170-year, ~208-year and ~350-year solar cycles. As sunspot counts during parts of the Maunder Minimum remained at zero with little fluctuation, it is reasonable to exclude the dates before 1705 in the comparison of the model to the data. Although solar activity may have still fluctuated during the Maunder Minimum, it may have remained below a threshold level that allows sunspots to occur. This would make sunspot counts a poor indicator for planetary tidal activity during the Maunder Minimum. Other than that, the model is able to reproduce the majority of features seen in the moving average. The Dalton Minimum (1780-1840) is accounted for by the ~170-year solar cycle's second last sunspot minimum. The lull in solar activity around the turn of the 20th century is accounted for by the last sunspot minimum of the ~208-year solar cycle. The three hump-like shapes occurring in the latter halves of the 18th, 19th and 20th centuries respectively roughly overlap the first, second and third industrial revolutions outlined in figure 9K. The strong correlation of this model with the data justifies the approach of approximating the sunspot record as a superposition of repeating negative delta functions.

name	amplitude	period	Breadth	phase shift
Gleissberg cycle	1	89.5	30	2020.5
170-year cycle	0.5	170	15	1980
Seuss cycle	1	208	30	1893
350-year cycle	1	350	15	1686
Millennial-scale cycle	1	895.5	300	1486
Hallstatt cycle	2	2400	300	1486

Figure 9R: Summary of solar periodicities used in this composite model of long-term solar variability. The amplitudes are estimated from figure 9C's spectral curve. All other quantities are measured in years.

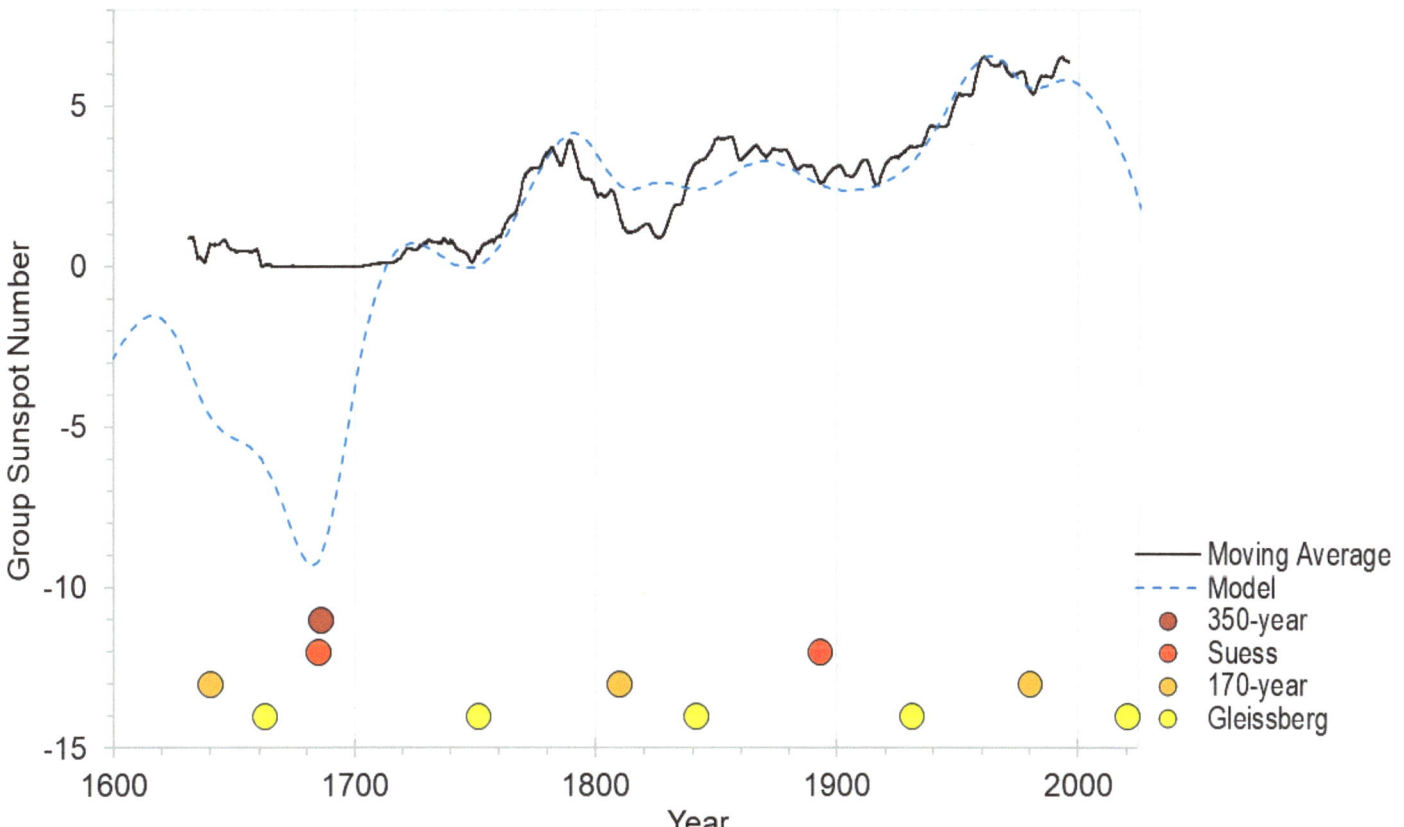

Figure 9S: The blue dashed line is the composite solar variability model generated from figure 9R. The black curve is figure 9G's moving average of monthly group sunspot numbers. The coloured circles represent sunspot minima of individual periodicities.

Discussion and Conclusion

A potential link between planetary alignments and cycles in solar activity has been dismissed by many authors in the past based on accusations of poor sampling methodology and failed predictions. Theories related to planetary influences on solar periodicities also don't benefit from a lack of consensus among astronomers and physicists about the nature and timing of long-period solar cycles. The very existence of solar periodicities longer than the ~11-year solar cycle has been disputed as recently as 2019 for not being significantly above random noise in the sunspot record (Cameron and Schuessler 2019). Yet despite existing somewhat on the fringes of the scientific mainstream for decades, modern astronomers still have a hard time completely ignoring a potential connection between planetary alignments and solar activity.

Although the physical mechanism for a connection between planetary alignments and solar activity is not exactly known, several potential explanations have been proposed. One is that an increased tidal influence on the Sun's surface could disrupt the activity of the Sun's surface magnetic field, quietly lowering the Sun's total magnetic activity and lowering overall solar output. Another potential explanation is that planetary tidal torques affect the position of the tachocline, setting off chain reactions that affect the Sun's nuclear furnace. However it happens, the combined tidal torque of all of the planets on the Sun's surface layer has been calculated to produce variations on the millimeter scale (Pikovsky et al 2001). Therefore, it is necessary to have an amplification process where these minor changes resonate harmoniously over long periods of time. Yet the truth is that it might not be entirely necessary to identify all aspects of the physical mechanism in order to show that a link between planetary alignments and solar activity exists.

In fact, the absence of a sufficient alternative explanation for quasi-periodic solar oscillations justifies the assertion that solar variability could be at least partially linked to the gravitational influence of the planets. Since patterns in solar activity are shown to be somewhat periodic, explanations based solely on the Sun's internal dynamics are unfavourable. In reality, the different solar periodicities intertwine alongside occasional stochastic events which contribute to the evolution of solar activity.

The quasi-periodic nature of solar cycles also reflects the fact that each time a set of planets align, the conjunction doesn't necessarily have the same degree of precision. Therefore, each planetary-induced tidal maximum can diverge from the expected date predicted by the cycle's average length to some degree. This minor level of inconsistency contributes to past solar activity patterns being mostly periodic with some non-periodic influences. As the quasi-periodic nature of solar cycles makes it difficult to predict their maxima and minima with 100% accuracy, a comparison between expected dates of particular types of planetary alignments and historical events should not be prematurely dismissed if they do not match precisely. Just think of the weather. The middle of January is not always colder than the middle of February, but we can expect it to be for any set of years that are measured. Therefore, solar activity patterns need not exactly match the simple mathematical model presented in this article in order for planetary alignments to still have a genuine influence on solar activity.

This study analyzed cyclical patterns of solar activity and their relation to some important planetary alignments during the latter part of the Holocene. First, the historical and isotopic sunspot records were decomposed into a series of overlapping periodicities. The maxima and minima of several periodicities were then compared with certain types of planetary alignments and historical events that have a repetitive nature. Several lines of evidence were presented to suggest that planetary alignments have a noticeable effect on historical events of environmental, economic and political importance. Historically significant planetary alignments were illustrated in figures 9F, 9I, 9N and 9P. Expected dates for particular types of planetary alignments were tabulated in figures 9J, 9K, 9L, 9M, 9O and 9Q alongside selected historical events. Through this analysis, it was deduced that times of maximal planetary tidal torque tend to correspond with times of reduced solar output, colder temperatures on Earth, and reduced economic prosperity.

The influence of solar periodicities might be useful for describing general patterns in the historical record. In this study, a mathematical model for sunspot variation was developed and compared to the historical sunspot record from 1705 to 1995, which yielded a Pearson correlation coefficient of 93%.

Although this analysis merely scratched the surface at times, the rough correlations between figure 9E, 9I, 9N and 9P's planetary alignments and historically significant economic and political events are difficult to dismiss as random coincidences.

Although an exact physical mechanism connecting planetary tidal influence to increases and decreases in solar output is not exactly known, a number of interesting relationships between commonly reported solar periodicities and historical events should be noted. An ~11-year solar cycle may have an influence on agriculture and the economy, although that claim has been contested for a few centuries. The rough correlation between sunspot minima of the ~11-year solar cycle and periods of economic hardship such as the 1970s energy crisis, Black Monday of 1987, the 1997 Asian financial crisis, the 2008 financial crisis and the 2020 stock market crash is intriguing.

A solar periodicity of 80 to 100 years called the Gleissberg cycle has also been identified. This study explored a potential connection between the Gleissberg cycle and a planetary alignment pattern involving conjunctions of the six innermost planets with a repeating period of ~89.5 years. The last two tidal maxima in this pattern, January 1931 and July 2020, roughly correspond with two historical economic depression events that stand out: the Great Depression and the recession following 2008. The last three tidal minima of this ~89.5-year pattern generally correspond with times of economic prosperity, taking place ±6.25 years from the centers of the first, second and third industrial revolutions.

This article has also shown that the most recent sunspot minima of the ~170-year, ~208-year and ~350-year oscillations in solar activity have roughly coincided with times of economic depression. The last three sunspot minima of the ~170-year solar cycle were associated with the beginning of the Maunder Minimum (1645-1715), the middle of the Dalton Minimum (1780-1840) and a global cooling episode in the 1960s and 1970s. The last three sunspot minima of the ~208-year solar cycle overlapped the Sporer Minimum (1460-1555), the Maunder Minimum, and a period of decreased economic activity known as the Long Depression (1873-1896). The last two sunspot minima of the ~350-year solar cycle overlapped the Wolf Minimum (1280-1350) and the Maunder Minimum. The ~170-year and ~350-year solar cycles are probably related to beats and harmonics of other planetary alignment patterns. On the other hand, the ~208-

year solar cycle is probably based more on the dynamics of the Earth-Moon system and its relationship to the Sun and stars. The sunspot minima of the ~170-year, ~208-year and ~350-year solar cycles superimposed during the Maunder Minimum, which was the nadir of the Little Ice Age.

A commonly reported solar periodicity based on a timescale of 900-1000 years, sometimes called the Eddy cycle, was simply referred to as the millennial-scale solar cycle in this article. Here, it was shown that it might be linked to a pattern of close conjunctions involving Jupiter, Saturn and Neptune with a repeating period of ~895.5 years. This solar periodicity may be responsible for large-scale oscillations in Earth's climate that produced the Roman Warm Period, the Medieval Warm Period and the Modern Warm Period. In turn, it may have been partially responsible for major patterns in agriculture, settlement, trade, migration and state dynamics among human societies during the Holocene.

The last solar periodicity described here was a ~2400-year oscillation known as the Hallstatt cycle. This cycle is probably based on the Sun's inertial motion with respect to the solar system's barycenter. Among solar periodicities longer than 11 years, the Hallstatt cycle has the largest amplitude. It may have been the primary cause of certain historical cool periods including the Little Ice Age (1300-1850) and the Hallstattzeit cool period (750-400 BC). Tidal maxima of the ~2400-year cycle have also roughly coincided with some important times for human civilization including the Axial Age and the Renaissance.

The mathematical model developed in this study to approximate the historical sunspot record used a composition of overlapping periodic negative delta functions of the form

$$S_P(t) = -\sum_{i=1}^{\infty} A e^{-\left(\frac{t-(\phi \pm iP)}{B}\right)^2}$$

These functions reflect the nature of periodic tidal pulses on the Sun by planets more accurately than sinusoidal functions and greatly enhance the ability to mimic the historical sunspot record. The model in this article generates a strong correlation between the calculated planetary tidal influence and multi-decadal trends in the historical sunspot record over the last few centuries. This model is less accurate during the Maunder Minimum due to a flatline in sunspot counts.

Among the commonly reported solar periodicities longer than 11-years, the Gleissberg cycle and the millennial-scale solar cycle stand out as all four phases of them (minimum, waxing, maximum, waning) have an influence on historical events whereas the effects of most other periodicities are only identifiable near their tidal maxima. The reason for this distinction is unclear but it's possible that solar periodicities related to the perihelia of Jupiter and Saturn behave more like sinusoidal patterns whereas other periodicities only manifest as concentrated sunspot minima separated by lengthy periods of pseudo-equilibria. Yet, it may also be the case that the Gleissberg cycle and the millennial-scale solar cycle resonate with sociological patterns more than the other solar periodicities.

Although this study provided insight into many aspects of solar periodicities and related planetary alignment patterns, several assumptions were made and many questions about the topic remain unanswered. The study was not able to further understanding of physical processes that could translate planetary tidal torques into changes in solar output. It did not explain quantitative relationships between changes in the planetary tidal torque and changes in sunspot counts beyond the general assertion that a larger tidal influence promotes a quieter Sun and a slightly reduced solar output. This study also did not provide any quantitative link between changes in the number of sunspots and changes in Earth's average temperature or indicators of economic prosperity. Instead, it focused more on qualitative relationships between variables and relied on the general assumption that a more active solar magnetic field leads to a warmer Earth and increased economic prosperity. Such qualitative relationships were assumed to be simple linear relationships and applied as such while processing data and generating the mathematical model of past solar activity. This approach is justified, however, as exploring the validity of these assumptions would have greatly expanded the size of this study and distracted from its main focus.

The implications of this study's results are numerous and extensive. The results of this study may help shed light on how the Sun behaves and why Earth's climate oscillates on a variety of timescales. If planetary alignments genuinely affect sunspot counts, then long-term trends in solar activity can be predicted and more information related to the nature of sunspots can be revealed. This in turn could contribute to a more in-depth understanding of long-term trends in economic and sociological indicators in the historical record.

The potential influence of certain types of planetary alignments on solar activity, climate and economic prosperity conjectured here was supported by novel methods and relatively simple mathematical formulations. This study revealed that a number of important historical events may be related to special planetary arrangements. Like astrologers in ages past suggested, the orbit of the planets may affect what happens on Earth.

Acknowledgments

Bill Howell is thanked for a number of discussions that helped refine this article.

References

Abreu,J.A.,Beer,J.,Ferriz-Mas,A.,McCracken,K.G.,Steinhilber, F.(2012).Is there a planetary influence on solar activity? *Astronomy & Astrophysics,548,*A88.

Anet,J.G.,Muthers,S.,Rozanov,E.V.,Raible,C.C.,Stenke,A., Shapiro,A.I.,Peter,T.(2014).Impact of solar versus volcanic activity variations on tropospheric temperatures and precipitation during the Dalton Minimum.*Climate of the Past, 10*(3),921-938.

Babcock,H.W.(1961).The Topology of the Sun's Magnetic Field and the 22-YEAR Cycle.*Astrophysical Journal,vol. 133,*572.

Baidolda,F.(2017).Search for planetary influences on solar activity(Doctoral dissertation,Université Paris sciences et lettres).

Biswas,A.,Karak,B.B.,Usoskin,I.,Weisshaar,E.(2023).Long-term modulation of solar cycles.*Space Science Reviews, 219*(3),19.

Breitenmoser,P.,Beer,J.,Brönnimann,S.,Frank,D.,Steinhilber, F.,Wanner,H.(2012).Solar and volcanic fingerprints in tree-ring chronologies over the past 2000 years.*Palaeogeography,Palaeoclimatology,Palaeoecology,313,*127-139.

Büntgen,U.,Myglan,V.S.,Ljungqvist,F.C.,McCormick,M.,Di Cosmo,N.,Sigl,M.,Kirdyanov,A.V.(2016).Cooling and societal change during the Late Antique Little Ice Age from 536 to around 660 AD.*Nature geoscience,9*(3),231-236.

Cameron,R.H.,Schuessler,M.(2019).Solar activity: periodicities beyond 11 years are consistent with random forcing. *Astronomy & Astrophysics,625,*A28.

Campbell,I.D.,Campbell,C.,Apps,M.J.,Rutter,N.W.,Bush,A. (1998).Late Holocene similar to 1500yr climatic periodicities and their implications.*Geology.*26:471–473.

Charbonneau,P.(2020).Dynamo models of the solar cycle.*Living Reviews in Solar Physics,17*(1),1-104.

Conway,E.(2008).What's in a name?Global warming vs.climate change.*Washington,DC:National Aeronautics and Space Administration (NASA),Internet Resource.*

Crafts,N.(1994).The industrial revolution.*The Economic History of Britain Since 1700,1,*1-16.

Damon,P.E.,Sonett,C.P.(1991).Solar and terrestrial components of the atmospheric ^{14}C variation spectrum.*The sun in time,*360.

Davis,O.K.,Jirikowic,J.,Kalin,R.M.(1992).Radiocarbon Record of Solar Variability and Holocene Climatic Change.In *Proceedings of the Seventh Annual Pacific Climate (PACLIM) Workshop:Asilomar,California,March 10-13,1991*(19).California Department of Water Resources,Interagency Ecological Studies Program for the Sacramento-San Joaquin Estuary.

Eddy,J.A.(1976).The Maunder Minimum:The reign of Louis XIV appears to have been a time of real anomaly in the behavior of the sun.*Science,192*(4245),1189-1202.

Fahrni,S.M.,Southon,J.,Fuller,B.T.,Park,J.,Friedrich,M., Muscheler,R.,Taylor,R.E.(2020).Single-year German oak and Californian bristlecone pine ^{14}C data at the beginning of the Hallstatt plateau from 856 BC to 626 BC.*Radiocarbon, 62*(4),919-937.

Fletcher,T.W.(2013).The great depression of English agriculture,1873–1896.In *British Agriculture*(30-55). Routledge.

Glasner,D.(2013).Business cycles and depressions:An encyclopedia.*Routledge*

Gorbanev,M.(2012).Sunspots,unemployment,and recessions, or Can the solar activity cycle shape the business cycle?.

Hathaway,D.H.(2015).The solar cycle.*Living reviews in solar physics,12*(1),1-87.

Herschel,W.(1801).Observations tending to investigate the nature of the sun, in order to find the causes or symptoms of its variable emission of light and heat; with remarks on the use that may possibly be drawn from solar observations.*Philosophical Transactions of the Royal Society of London,*(91),265-318.

Hoyt,D.V.,& Schatten,K.H.(1998).Group sunspot numbers:A new solar activity reconstruction.*Solar physics,179*(1),189-219.

James,P.,Thorpe,I.J.(1991).Centuries of darkness:a challenge to the conventional chronology of Old World archaeology.*Rutgers University Press*.

Jones,E.T.,Hewlett,R.,Mackay,A.W.(2021).Weird weather in Bristol during the Grindelwald Fluctuation(1560–1630).*Weather*,76(4),104-110.

Kopecky,M.(1991).When did the latest minimum of the 80-year sunspot period occur?.*Bulletin of the Astronomical Institutes of Czechoslovakia*,42,158-160.

Korschinek,G.,Bergmaier,A.,Faestermann,T.,Gerstmann,U.C.,Knie,K.,Rugel,G.,Remmert,A.(2010).A new value for the half-life of ^{10}Be by heavy-ion elastic recoil detection and liquid scintillation counting.*Nuclear Instruments and Methods in Physics Research Section B:Beam Interactions with Materials and Atoms*,268(2),187-191.

Kotzé,P.B.(2020).Identification of solar periodicities in southern African baobab δ^{13}C record.*South African Journal of Science*,116(7-8),1-5.

Kotzé,P.(2023).Variability of solar periodicities in southern African baobab δ^{13}C data during the Wolf,Spörer,Maunder and Dalton minima.*Journal of Atmospheric and Solar-Terrestrial Physics*,247,106075.

Kronig,O.,Ivy-Ochs,S.,Hajdas,I.,Christl,M.,Wirsig,C.,Schlüchter,C.(2018).Holocene evolution of the Triftje-and the Oberseegletscher (Swiss Alps) constrained with ^{10}Be exposure and radiocarbon dating.*Swiss Journal of Geosciences*,111,117-131.

Lamb,H.H.(1977).Climate.Vol.2

Love,J.J.(2013).On the insignificance of Herschel's sunspot correlation.*Geophysical Research Letters*.40(16):4171-4176.

Ma,L.,Yin,Z.,Han,Y.(2018).Quasi~ 500-year Cycle Signals in Solar Activity.*Earth Sci.Res*,7,131-136.

Mann,M.E.,Zhang,Z.,Rutherford,S.,Bradley,R.S.,Hughes,M.K.,Shindell,D.,Ni,F.(2009).Global signatures and dynamical origins of the Little Ice Age and Medieval Climate Anomaly.*Science*,326(5957),1256-1260.

Martin-Puertas,C.,Matthes,K.,Brauer,A.,Muscheler,R.,Hansen,F.,Petrick,C.,Van Geel,B.(2012).Regional atmospheric circulation shifts induced by a grand solar minimum.*Nature Geoscience*,5(6),397-401.

Matthews,J.A.,Briffa,K.R.(2005).The'Little Ice Age':re-evaluation of an evolving concept.*Geografiska Annaler:Series A,Physical Geography*,87(1),17-36.

McCracken,K.G.,Beer,J.,Steinhilber,F.,Abreu,J.(2013).A phenomenological study of the cosmic ray variations over the past 9400 years,and their implications regarding solar activity and the solar dynamo.*Solar Physics*,286,609-627.

Mitchell Jr,J.M.(1963).On the world-wide pattern of secular temperature change.In *Changes of Climate*(Vol.20,161-181).Unesco Paris.

Mohajan,H.(2021).Third industrial revolution brings global development.

Mokyr,J.,Strotz,R.H.(1998).The second industrial revolution,1870-1914.*Storia dell'economia Mondiale*,21945(1).

Murray,C.D.,Dermott,S.F.(1999).Solar system dynamics.*Cambridge university press*.

National Oceanic and Atmospheric Administration.(2022).ISES Solar Cycle Sunspot Number Progression.Space Weather Prediction Center(website).

Paluš,M.,Kurths,J.,Schwarz,U.,Seehafer,N.,Novotná,D.,Charvátová,I.(2007).The solar activity cycle is weakly synchronized with the solar inertial motion.*Physics Letters A*,365(5-6),421-428.

Parker,E.N.(1958).Dynamics of the interplanetary gas and magnetic fields.*The Astrophysical Journal*,128,664.

Pikovsky,A.,Rosenblum,M.,Kurths,J.(2001).Synchronization:A universal concept in nonlinear sciences.*Self*,2,3.

Poluianov,S.V.,Kovaltsov,G.A.,Mishev,A.L.,Usoskin,I.G.(2016).Production of cosmogenic isotopes ^{7}Be,^{10}Be,^{14}C,^{22}Na,and ^{36}Cl in the atmosphere:Altitudinal profiles of yield functions.*Journal of Geophysical Research:Atmospheres*,121(13),8125-8136.

Poynting,J.H.(1884).A comparison of the fluctuations in the price of wheat and in the cotton and silk imports into Great Britain.*Journal of the Statistical Society of London*,47(1),34-74.

Robles,M.,Peyron,O.,Brugiapaglia,E.,Ménot,G.,Dugerdil,L.,Ollivier,V.,Joannin,S.(2022).Impact of climate changes on vegetation and human societies during the Holocene in the South Caucasus (Vanevan,Armenia):A multiproxy approach including pollen, NPPs and brGDGTs.*Quaternary Science Reviews*,277,107297.

Ruzmaikin,A.,Feynman,J.,Yung,Y.L.(2006).Is solar variability reflected in the Nile River?*Journal of Geophysical Research:Atmospheres*,111(D21).

Scafetta,N.(2014).Discussion on the spectral coherence between planetary, solar and climate oscillations: a reply to some critiques.*Astrophysics and Space Science*,354(2),275-299.

Scafetta,N.,Milani,F.,Bianchini,A.,Ortolani,S.(2016).On the astronomical origin of the Hallstatt oscillation found in radiocarbon and climate records throughout the Holocene.*Earth-Science Reviews*,162,24-43.

Scafetta,N.,Bianchini,A.(2022).The planetary theory of solar activity variability:a review.*Frontiers in Astronomy and Space Sciences*,9,937930.

Schurer,A.P.,Hegerl,G.C.,Luterbacher,J.,Brönnimann,S.,Cowan,T.,Tett,S.F.,Timmreck,C.(2019).Disentangling the causes of the 1816 European year without a summer.*Environmental Research Letters*,14(9),094019.

Solanki,S.K.(2003).Sunspots:an overview.*The Astronomy and Astrophysics Review*,11(2),153-286.

Solanki,S.K.,Usoskin,I.G.,Kromer,B.,Schüssler,M.,& Beer,J. (2004).Unusual activity of the Sun during recent decades compared to the previous 11,000 years.*Nature*,*431*(7012), 1084-1087.

Sonett,C.P.,Giampapa,M.S.,Matthews,M.S.(Eds.).(1991).The sun in time.*University of Arizona Press*.

Spiegel,E.A.,Weiss,N.O.(1980).Magnetic activity and variations in solar luminosity.*Nature*,*287*(5783),616-617.

Stefani,F.,Giesecke,A.,Weber,N.,Weier,T.(2016).Synchronized helicity oscillations:A link between planetary tides and the solar cycle?*Solar Physics*,*291*(8),2197-2212.

Stefani,F.,Giesecke,A.,Weier,T.(2019).A model of a tidally synchronized solar dynamo.*Solar Physics*,*294*(5),60.

Steinhilber,F.,Abreu,J.A.,Beer,J.,Brunner,I.,Christl,M., Fischer,H.,Wilhelms,F.(2012).9,400 years of cosmic radiation and solar activity from ice cores and tree rings.*Proceedings of the National Academy of Sciences*,*109*(16),5967-5971.

Strauss,W.,Howe,N.(1991).Generations:The history of America's future,1584 to 2069.*(No Title)*.

Stuiver,M., Quay,P.D.(1980).Changes in atmospheric carbon-14 attributed to a variable sun.*Science*,*207*(4426),11-19.

Usoskin,I.G.,Arlt,R.,Asvestari,E.,Hawkins,E.,Käpylä,M.,Kovaltsov,G.A.,Vaquero,J.M.(2015).The Maunder minimum (1645-1715) was indeed a grand minimum: A reassessment of multiple datasets.*Astronomy & Astrophysics*,*581*, A95.

Usoskin,I.G.,Gallet,Y.,Lopes,F.,Kovaltsov,G.A.,Hulot,G. (2016).Solar activity during the Holocene: the Hallstatt cycle and its consequence for grand minima and maxima.*Astronomy & Astrophysics*,*587*,A150.

Usoskin,I.G.(2017).A history of solar activity over millennia.*Living Reviews in Solar Physics*,*14*(1),3.

Van der Plicht,J.(2005).Radiocarbon,the calibration curve and Scythian chronology.In *Impact of the Environment on Human Migration in Eurasia* (45-61).Springer Netherlands.

Vokhmyanin,M.,Arlt,R.,Zolotova,N.(2020).Sunspot positions and areas from observations by Thomas Harriot.*Solar Physics*,*295*(3),1-11.

Wagner,S.,Zorita,E.(2005).The influence of volcanic,solar and CO2 forcing on the temperatures in the Dalton Minimum (1790–1830):A model study *Climate dynamics*,*25*,205-218.

Weiss,N.O.,Tobias,S.M.(2016).Supermodulation of the Sun's magnetic activity:the effects of symmetry changes. *Monthly Notices of the Royal Astronomical Society*,*456*(3), 2654-2661.

Wickson,S.(2023)The 900-year climate cycle: an analysis of global events in the Holocene.ISBN:9798378565559.Independently published.

Wolf,R.(1859).Mittheilungen über die Sonnenflecken.*Druck von Zürcher und Furrer*.

Wolff,C.L.,Patrone,P.N.(2010).A new way that planets can affect the Sun. *Solar Physics*,*266*(2),227-246.

Wu,C.J.,Usoskin,I.G.,Krivova,N.,Kovaltsov,G.A.,Baroni,M.,Bard,E.,& Solanki,S.K.(2018).Solar activity over nine millennia:A consistent multi-proxy reconstruction.*Astronomy & Astrophysics*,*615*,A93.

Xu,D.,Lu,H.,Chu,G.,Wu,N.,Shen,C.,Wang,C.,Mao,L.(2014). 500-year climate cycles stacking of recent centennial warming documented in an East Asian pollen record.*Scientific Reports*,*4*(1),3611.

Zharkova,V.V.,Shepherd,S.J.,Popova,E.,Zharkov,S.I.(2015). Heartbeat of the Sun from Principal Component Analysis and prediction of solar activity on a millenium timescale. *Scientific reports*,*5*(1),15689.

Zharkova,V.,Shepherd,S.,Popova,E.,Zharkov,S.,Xia,Q.(2018, April).Upcoming modern grand minimum and solar activity prediction backwards five millennia.In *EGU General Assembly Conference Abstracts*(p.8066).

Zharkova,V.(2020).Modern Grand Solar Minimum will lead to terrestrial cooling.*Temperature*.

Zharkova,V.V.,Vasilieva,I.,Shepherd,S.J.,Popova,E.(2023). Periodicities in solar activity, solar radiation and their links with terrestrial environment.*Natural Science*,*15*(3),111-147.

Zhen-tao,X.(1980).The hexagram "feng" in "the book of changes" as the earliest written record of sunspot.*Chinese Astronomy*,*4*(4),406.